Nitrate in the Mississippi River and Its Tributaries, 1980–2010: An Update

By Jennifer C. Murphy, Robert M. Hirsch, and Lori A. Sprague

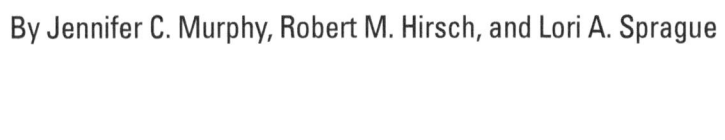

National Water-Quality Assessment Program

Scientific Investigations Report 2013–5169

U.S. Department of the Interior
U.S. Geological Survey

U.S. Department of the Interior
SALLY JEWELL, Secretary

U.S. Geological Survey
Suzette M. Kimball, Acting Director

U.S. Geological Survey, Reston, Virginia: 2013

For more information on the USGS—the Federal source for science about the Earth, its natural and living resources, natural hazards, and the environment, visit http://www.usgs.gov or call 1–888–ASK–USGS.

For an overview of USGS information products, including maps, imagery, and publications, visit http://www.usgs.gov/pubprod

To order this and other USGS information products, visit http://store.usgs.gov

Suggested citation:
Murphy, J.C., Hirsch, R.M., and Sprague, L.A., 2013, Nitrate in the Mississippi River and its tributaries, 1980–2010—An update: U.S. Geological Survey Scientific Investigations Report 2013–5169, 31 p., http://pubs.usgs.gov/sir/2013/5169/.

Contents

Figures

Tables

Conversion Factors

SI to Inch/Pound

Multiply	By	To obtain
Length		
meter (m)	3.281	foot (ft)
kilometer (km)	0.6214	mile (mi)
Area		
square meter (m^2)	10.76	square foot (ft^2)
square kilometer (km^2)	0.3861	square mile (mi^2)
Flow rate		
cubic meter per second (m^3/s)	35.31	cubic foot per second (ft^3/s)
Mass		
kilogram (kg)	2.205	pound avoirdupois (lb)

Concentrations of chemical constituents in water are given in milligrams per liter (mg/L).

Nitrate in the Mississippi River and Its Tributaries, 1980–2010: An Update

By Jennifer C. Murphy, Robert M. Hirsch, and Lori A. Sprague

Abstract

Nitrate concentration and flux were estimated from 1980 through 2010 at eight sites in the Mississippi River Basin as part of the National Water-Quality Assessment (NAWQA) Program of the U.S. Geological Survey (USGS). These estimates extend the results from a previous investigation that provided nitrate estimates from 1980 through 2008 at the same sites. From 1980 through 2010, annual flow-normalized (FN) nitrate concentration and flux in the Iowa and Illinois Rivers decreased by 11 to 15 percent. These two rivers had the highest FN nitrate concentration in 1980 (5.3 milligrams per liter [mg/L] and 3.9 mg/L, respectively) of any of the study sites. Nitrate increased in the Missouri River (79 and 45 percent increase in FN concentration and flux, respectively), and at the four sites on the Mississippi River (17 to 70 percent increase in FN concentration and 8 to 55 percent increase in FN flux) from 1980 through 2010. Nitrate in the Ohio River was generally stable during this time. Historically, nitrate was high and changed little in the Iowa and Illinois Rivers; however, nitrate concentrations began to decrease around 2000, and this decrease continued through 2010. Also during this time, near-flat nitrate trends in lower sections of the Mississippi River began increasing, likely reflecting the acceleration of already increasing nitrate trends in the upper Mississippi and Missouri Rivers, in addition to increases in inputs from other tributaries in the lower part of the Mississippi River Basin. Spring trends (April through June) generally parallel annual trends at all sites from 1980 through 2010, except in the Iowa River where decreasing nitrate during the spring was not observed. In general, most sites had increases in nitrate concentration at low streamflows, which suggests increases in legacy nitrate from groundwater or point source contributions. In aggregate, the decreases in nitrate concentrations from the Iowa and Illinois Rivers, which largely occurred during high flows, appear to be overshadowed by increasing nitrate concentrations across much of the Mississippi River Basin.

Introduction

Nitrate flux from the Mississippi River is a major determinant of the extent and severity of the hypoxic zone that forms in the northern Gulf of Mexico every summer (Scavia and others, 2003; Turner and others, 2006, 2012). This hypoxic zone is one of the largest in the world (Rabalais and others, 2002), and reducing its size is an important national goal (Mississippi River/Gulf of Mexico Watershed Nutrient Task Force, 2008). Accurate estimation of nitrate flux from the Mississippi River Basin (MRB) to the Gulf is a vital step towards this goal and has been a critical scientific task for over a decade (Goolsby and others, 2000; Goolsby and Battaglin, 2001; McIssac and others, 2001). A recent study by Sprague and others (2011) used a new approach, the "weighted regressions on time, discharge, and season" (WRTDS) method (Hirsch and others, 2010), to estimate nitrate plus nitrite (termed nitrate hereafter) concentration and flux from 1980 through 2008 at eight sites in the MRB (fig. 1). Sprague and others (2011) reported no consistent declines in nitrate concentration or flux at any of the sites. Furthermore, from 1980 through 2008 nitrate increased considerably in the upper Mississippi River and the Missouri River and remained the same or increased at a slower rate at the other sites (Sprague and others, 2011). This report follows that of Sprague and others (2011) and extends estimates of nitrate concentration and flux through 2010.

This report presents estimates of nitrate concentration and flux at eight sites in the MRB from 1980 through 2010 (table 1) and was completed as part of the National Water-Quality Assessment (NAWQA) Program of the U.S. Geological Survey (USGS). It is one of several studies that have been conducted across the Nation to provide a better understanding of changes in nutrient concentrations and fluxes over time. Using the methods described in Sprague and others (2011), this report documents (1) noteworthy developments in trends with the extension of estimates from 2008 (Sprague and others, 2011) through 2010, (2) 30-year, 20-year, and 10-year annual trends in nitrate during 1980–2010, 1980–2000, and 2000–2010, respectively, and (3) spring trends in nitrate (April through June) for the same periods. This report also examines the variability of expected nitrate concentrations in relation to season and streamflow at each of the eight sites for a recent period from 2000 through 2010.

Figure 1.   Mississippi River Basin and study sites, and a schematic line drawing of the relative locations of study sites, major tributaries, and additional sites used for streamflow data. Flux estimates at each site include entire upstream drainage basin.

Table 1.   Site and data characteristics of the eight Mississippi River Basin sites used in this study (1980–2010).

[km², square kilometers; USGS, U.S. Geological Survey]

Site abbreviation	Site number	Site name	Basin area (km²)	Model calibration		Number of samples
				Start date	End date	
MSSP-CL	05420500	Mississippi River at Clinton, IA	221,703	11/12/1974	8/16/2011	363
IOWA-WAP	05465500	Iowa River at Wapello, IA	32,375	11/10/1977	9/1/2011	333
ILLI-VC	05586100	Illinois River at Valley City, IL	69,264	12/12/1974	8/24/2011	428
MSSP-GR	05587455[1]	Mississippi River below Grafton, IL	443,665	1/27/1975	9/14/2011	362
MIZZ-HE	06934500	Missouri River at Hermann, MO	1,353,269	10/28/1969	9/29/2011	527
MSSP-TH	07022000	Mississippi River at Thebes, IL	1,847,180	1/30/1973	8/2/2011	511
OHIO-GRCH	03612500[2]	Ohio River at Dam 53 near Grand Chain, IL	526,027	10/11/1972	8/15/2011	466
MSSP-OUT	---[3]	Mississippi River above Old River Outflow Channel, LA	2,914,514	10/5/1967	8/22/2011	589

[1]Streamflow measured at Mississippi River at Grafton, IL (USGS site 05587450).

[2]Streamflow measured at Ohio River at Metropolis, IL (USGS site 03611500).

[3]MSSP-OUT is meant to provide an approximation of streamflow, concentration, and flux just upstream of the Old River Outflow Channel. Streamflow is the sum of Mississippi River at Tarbert Landing, MS (U.S. Army Corps of Engineers station 01100) and Old River Outflow Channel near Knox Landing, LA (U.S. Army Corps of Engineers station 02600), and nutrient data was sampled at Mississippi River near St. Francisville, LA (USGS site 07373420).

Methods

Following the methods of Sprague and others (2011), nitrate concentration and flux were estimated using the WRTDS method. WRTDS uses locally weighted regression to estimate daily nitrate concentration. Separate regression models are built for each day, resulting in unbiased estimates of concentration (Hirsch and others, 2010). In WRTDS, concentration is modeled as

$$\ln(c) = \beta_0 + \beta_1 t + \beta_2 \ln(Q) + \beta_3 \sin(2\pi t) + \beta_4 \cos(2\pi t) + \varepsilon \quad (1)$$

where

ln	is	the natural log;
c	is	nitrate concentration;
$\beta_0, \beta_1, \beta_2, \beta_3,$ and β_4	are	fitted coefficients;
t	is	time;
Q	is	mean daily streamflow; and
ε	is	unexplained variability.

The observations in the calibration dataset are weighted according to how close the values of time, season, and discharge lie to the respective observation made on the day being modeled. The tricube weight function (Tukey, 1977) is used to define the weights of the calibration dataset for each regression model. The half-window widths used in this study are the same as those used in Sprague and others (2011): 10 years for time, 0.25 for season, and 1 natural log cycle for sites with greater than (>) 250,000 square kilometers (km²) of drainage area or 2 natural log cycles for sites with less than or equal to (≤) 250,000 km² of drainage area for discharge. For details about WRTDS, see Hirsch and others (2010).

The methods used for this report depart from the computational methods used by Sprague and others (2011) in two ways—how censored observations are incorporated and how the bias adjustment is calculated. At the time Sprague and others (2011) conducted their study, WRTDS did not include the ability to evaluate censored data; therefore, Sprague and others (2011) set censored values to half the detection limit prior to analysis. Since 2011, WRTDS has been updated to incorporate censored values by using an adaptation of "survival analysis" (also known as "censored regression analysis"). Of the data used by Sprague and others (2011) and this report, less than 1 percent of the data were censored at the Missouri River site (MIZZ-HE), and all data were uncensored at the remaining sites. Also, in earlier versions of WRTDS, bias adjustment was accomplished using the Duan smearing estimator (Duan, 1983), whereas in this analysis, bias adjustment was based on the scale factor from the survival analysis. These changes are described in Moyer and others (2012, p. 9–11).

To check for the problem of flux estimation bias, the sum of the estimates of flux produced by WRTDS for each sampled day were compared to the sum of the actual fluxes determined from the sample. In no cases were the differences larger than 5 percent. This check for possible bias in flux estimates is similar to the procedure proposed by Stenback and others (2011) and used by Sprague and others (2011) and Garrett (2012).

Equation 1 is used to estimate concentration and flux for each day of the period of record; however, these estimates are strongly influenced by the discharge that occurred on those specific days. The flow normalization (FN) process in WRTDS takes the model of concentration for each day of the record (eq. 1) and integrates it over an estimated probability distribution of discharges for that particular day of the year. The set of observed discharges for any given day (in the case of this study, 31 values, one from each of the 31 years in the period of record) is used to represent the probability distribution for that day, with each of the 31 values considered to be equally likely. The FN concentration for a given day is the mean of the estimated concentrations for this day, integrated over this probability distribution of discharges. The FN flux for a given day is the mean of the estimated flux for this day (which is the estimated concentration multiplied by the associated discharge) integrated over this probability distribution of discharges. Thus, the FN concentration record is a representation of concentration after removing the influence of the year-to-year variations in discharge, and the FN flux record is a representation of flux after removing the influence of the year-to-year variations in discharge. FN estimates can be expected to give a more accurate representation of changes in the behavior of the watershed because the significant influence of random variations in discharge has been removed. For details about WRTDS and FN, see Hirsch and others (2010).

Daily nitrate concentration and flux (FN and non-FN) were estimated from 1980 through 2010 for the eight MRB sites reported in Sprague and others (2011) (fig. 1). Streamflow and nitrate data were compiled and prepared according to the methods described in Aulenbach and others (2007). The period of record for nitrate calibration datasets varied by site (table 1); streamflow datasets spanned from the start of the first water year (October 1 through September 30) to the end of the last water year of the nitrate record. Daily estimates were summarized into annual mean nitrate concentration and total annual nitrate flux by calendar year. These estimates, along with traces of annual FN concentration and flux, have been plotted for each station; an example for the upper Mississippi River (MSSP-CL) is shown in figure 2. Additionally, spring mean concentration and total spring flux are reported; spring concentration and flux only consider estimates from April through June (91-day period). For details pertaining to previous WRTDS modeling efforts at these sites, see Sprague and others (2011).

For each site, contour plots are used to visualize the expected value of nitrate concentration as a function of time, season, and streamflow from 2000 to 2010. Figure 3 depicts this surface for the upper Mississippi River (MSSP-CL). The thin black lines on this plot represent smoothed estimates of the 5th and 95th percentiles of streamflow and demonstrate the seasonal variability of streamflow. The vertical color changes on the graph show variation in expected nitrate concentration according to streamflow for any given day or time of year. Similarly, for a given streamflow, horizontal color changes show variation in expected nitrate concentration according to season and over time. Changes to expected nitrate concentration at low streamflows over time are largely attributable to changes in groundwater concentrations, although point source discharges could also be important sources in some cases. In contrast, changes in expected nitrate concentration at high streamflows over time are likely related to changes in the solute concentration of runoff during floods and storms.

Trends in Nitrate Concentration and Flux in the Mississippi River and Its Tributaries: An Update

As described by Sprague and others (2011), trends in FN nitrate concentration and flux were increasing or near-level at all sites from 1980 through 2008, but trends at select sites began to exhibit decreases (IOWA-WAP and ILLI-VC) or greater increases (MSSP-CL and MIZZ-HE) from 2000 through 2008. The extension of FN estimates through 2010 strengthens evidence for these nitrate trends that were beginning to develop earlier in the 2001–2010 decade. The Iowa River (IOWA-WAP) and Illinois River (ILLI-VC) showed small percentage increases or decreases in FN concentration and flux from 1980 through 2008 (Sprague and others, 2011); however, decreases of 11–15 percent at IOWA-WAP and ILLI-VC are now evident when estimates are extended through 2010 (table 2). The strengthening of FN nitrate trends with 2 years of data indicates that conditions in these basins are undergoing change, suggesting that management practices undertaken in the last decade or so may be having the effect of reducing nitrate fluxes (U.S. Department of Agriculture, Natural Resources Conservation Service, 2012). Also, the acceleration of increasing FN nitrate in the upper Mississippi River (MSSP-CL) and Missouri River (MIZZ-HE) observed from 2000 through 2008 (Sprague and others, 2011) continues to occur at a similar magnitude with the addition of 2 years of data at both sites (table 2).

Annual Trends

The 30-Year Annual Trends: 1980–2010

The largest increases in annual FN nitrate concentration and flux from 1980 through 2010 occurred in the upper Mississippi River (MSSP-CL) and Missouri River (MIZZ-HE) (table 2). FN nitrate concentration increased by 0.84 milligram per liter (mg/L) (70 percent), and flux increased by 39×10^6 kilograms per year (kg/yr) (55 percent) at MSSP-CL. Similarly, FN nitrate concentration at MIZZ-HE increased by 0.78 mg/L (79 percent), and flux increased by 42×10^6 kg/yr (45 percent). Along with the Ohio River site (OHIO-GRCH), these two sites had the lowest annual mean FN nitrate concentrations in 1980. FN nitrate concentration at OHIO-GRCH has remained low and varied little

over time (–0.02 mg/L or –2 percent change in FN concentration, and $–16 \times 10^6$ kg/yr or –5 percent change in FN flux). The highest FN nitrate concentrations in 1980 occurred in the Iowa River (IOWA-WAP) and Illinois River (ILLI-VC). These are the only two sites to have substantial decreasing trends in FN nitrate concentration during 1980–2010 (–0.60 mg/L and –0.55 mg/L, or –11 and –14 percent, respectively). FN fluxes at these sites also declined by about 15 percent during this period. The remaining sites (MSSP-GR, MSSP-TH, and MSSP-OUT) have moderate increases of 17 to 19 percent in FN nitrate concentration and increases of 8 to 14 percent for FN flux. Annual estimates of nitrate concentration and flux (FN and non-FN) from 1980 through 2010 are listed in appendix 1.

The 20-Year and 10-Year Trends: 1980–2000 and 2000–2010

Annual FN nitrate concentration and flux are either increasing or showed minimal change (minimal change is defined here as a change of less than ±10 percent) at all sites for the 20-year trend from 1980 through 2000 (table 3). Increases in FN nitrate concentration and flux occurred in upper (MSSP-CL) and middle (MSSP-GR) sections of the Mississippi River and the Missouri River (MIZZ-HE). FN nitrate flux in the Illinois River (ILLI-VC) also increased during this period. FN nitrate at the remaining sites exhibited minimal change (IOWA-WAP, MSSP-TH, OHIO-GRCH, and MSSP-OUT).

Recent trends (2000–2010) in annual FN nitrate were different from earlier trends (1980–2000) at several sites in the MRB (table 3). In the Iowa River (IOWA-WAP) and Illinois River (ILLI-VC), near-flat or increasing previous trends transitioned to decreasing trends around 2000. At these sites, recent trends in FN nitrate concentration were –10 and –21 percent, and recent trends in FN flux were –13 and –25 percent, respectively. Downstream of the Illinois River, recent (2000–2010) FN concentration and flux exhibited minimal change in the Mississippi River (MSSP-GR), reflecting a balance of different nitrate inputs from upstream sites such as MSSP-CL in the upper Mississippi River (increasing nitrate) and IOWA-WAP and ILLI-VC (decreasing nitrate). FN nitrate concentration and flux in the Ohio River (OHIO-GRCH) changed little during 2000–2010. The remaining MRB sites (MSSP-CL, MIZZ-HE, MSSP-TH, and MSSP-OUT) have increasing trends in FN nitrate concentration from 2000 through 2010. Positive recent trends in the upper Mississippi River (MSSP-CL) and Missouri River (MIZZ-HE) are consistent in direction with previous 20-year trends but are generally steeper in slope. In contrast, the slightly increasing trends (10–13 percent) at the two most downstream sites on the Mississippi River (MSSP-TH and MSSP-OUT) are new developments since about 2000 (table 3) and reflect increasing nitrate trends upstream.

Table 2.   The 30-year (1980–2010) annual and spring trends in flow-normalized nitrate concentration and flux at eight sites in the Mississippi River Basin.

[conc, nitrate concentration; FN, flow normalized; mg/L, milligrams per liter; kg/yr, kilograms per year; kg/km²/yr, kilogram per square kilometer per year; Increasing and decreasing trends are greater than or equal to ±10 percent, respectively, strongly increasing trends are greater 40 percent, and trends with minimal change are within ±10 percent. Spring trends and values consider a 91-day period from April through June]

Site	in 1980			30-year annual trends (1980–2010)			
				FN concentration		FN flux	
	FN annual mean conc (mg/L)	FN annual flux (10⁶ kg/yr)	FN annual yield (kg/km²/yr)	Change (mg/L)	Trend (percent change)	Change (10⁶ kg/yr)	Trend (percent change)
MSSP-CL	1.19	71	321	0.84	Strongly increasing (70)	39	Strongly increasing (55)
IOWA-WAP	5.28	63	1,955	−0.60	Decreasing (−11)	−9	Decreasing (−15)
ILLI-VC	3.85	102	1,478	−0.55	Decreasing (−14)	−15	Decreasing (−14)
MSSP-GR	2.59	342	771	0.45	Increasing (17)	39	Increasing (11)
MIZZ-HE	0.98	95	70	0.78	Strongly increasing (79)	42	Strongly increasing (45)
MSSP-TH	1.96	496	269	0.38	Increasing (19)	40	Minimal change (8)
OHIO-GRCH	0.99	310	589	−0.02	Minimal change (−2)	−16	Minimal change (−5)
MSSP-OUT	1.26	819	281	0.22	Increasing (17)	119	Increasing (14)

Site	in 1980		30-year spring trends (1980–2010)			
			FN concentration		FN flux	
	FN spring mean conc (mg/L)	FN spring flux (10⁶ kg/yr)	Change (mg/L)	Trend (percent change)	Change (10⁶ kg/yr)	Trend (percent change)
MSSP-CL	1.25	30	0.96	Strongly increasing (76)	17	Strongly increasing (56)
IOWA-WAP	6.00	27	0.43	Minimal change (7)	−0.3	Minimal change (−1)
ILLI-VC	4.89	44	−0.75	Decreasing (−15)	−7	Decreasing (−15)
MSSP-GR	3.19	146	0.72	Increasing (23)	27	Increasing (18)
MIZZ-HE	1.32	41	0.78	Strongly increasing (59)	14	Increasing (33)
MSSP-TH	2.54	221	0.43	Increasing (17)	12	Minimal change (6)
OHIO-GRCH	1.12	94	0.07	Minimal change (6)	0.7	Minimal change (1)
MSSP-OUT	1.54	325	0.38	Increasing (25)	54	Increasing (17)

Changes in Relative Contributions between 1980 and 2010

Between 1980 and 2010, FN nitrate flux changed by 119×10^6 kg/yr at the outflow of the Mississippi River (MSSP-OUT, table 2), and this mass can be proportionally attributed to the various nested upstream basin areas. The change in FN flux in the intervening subbasin above the Mississippi River at Grafton (MSSP-GR minus MSSP-CL, IOWA-WAP, and ILLI-VC) was 24×10^6 kg/yr, the change in FN flux in the intervening subbasin above the Mississippi River at Thebes (MSSP-TH minus MIZZ-HE and MSSP-GR) was -41×10^6 kg/yr, and the change in FN flux in the intervening subbasin above the outlet of the Mississippi River (MSSP-OUT minus MSSP-TH and OHIO-GRCH) was 95×10^6 kg/yr. The pattern of these changes in FN flux along the Mississippi River suggests increases at the outflow (MSSP-OUT) are primarily driven by increases in the intervening subbasin between the outflow and the confluence of the Ohio (OHIO-GRCH) and Mississippi Rivers, though increases in FN flux from the upper Mississippi River (MSSP-CL) and Missouri River (MIZZ-HE) also appear to be important. Increases in FN flux from the intervening subbasin above the outflow appear to be more than twice as large as contributions from any other intervening subbasin or tributary basin, effectively offsetting decreases in FN flux from the Iowa River (IOWA-WAP) and Illinois River (ILLI-VC). Interestingly, increases in FN flux of about 40×10^6 kg/yr from the Missouri River (MIZZ-HE) and immediately upstream in the Mississippi River (MSSP-GR) do not appear to result in a substantial change of FN flux immediately downstream (MSSP-TH), as would be expected. As reported by Sprague and others (2011), the decrease in the intervening subbasin above MSSP-TH may be caused by a variety of factors, such as a decrease in point source input from an upstream wastewater treatment plant that was upgraded around 1992, though increases in conservation efforts in this area may also play a role. The analysis of relative contributions is important to understanding the flux of nitrate through the MRB, however, estimates of flux contributions from intervening subbbasins may be affected by issues such as changes in in-stream processing of nitrogen, changing input from small tributaries and point sources, and the propagation of errors by differencing flux records from upstream and downstream sites.

Spring Trends

The portion of annual nitrate flux that occurs during the spring each year is particularly important because of its ability to stimulate spring and summer algae blooms, which directly influence the hypoxic zone in the Gulf of Mexico (Scavia and others, 2003). Therefore, in addition to the aggregation of nitrate concentration and flux estimates annually, estimates were aggregated over a 91-day spring period from April through June during 1980–2010 (appendix 2). Spring trends of FN nitrate concentration and flux from 1980 through 2010 closely parallel the annual trends at these sites for this same period, except in the Iowa River (IOWA-WAP) (table 2). At IOWA-WAP, FN concentration and flux exhibited minimal change during the spring (7 percent and –1 percent, respectively), as compared to decreasing nitrate trends annually over this period. The decoupling of annual and spring trends at IOWA-WAP suggests that even though FN nitrate may not have changed much during the spring, decreases during other times of the year are substantial enough to reduce FN nitrate in the Iowa River annually.

Historically, during 1980–2000, FN nitrate concentration and flux increased or were stable during the spring at all sites (table 4), similar to annual trends; however, increases in FN nitrate concentration and flux occurred during the spring at several sites even though annual trends at these sites changed little (IOWA-WAP, OHIO-GRCH, MSSP-OUT, see tables 3 and 4). This finding suggests increases in FN nitrate concentration or flux during the spring were not substantial enough to influence annual mean nitrate concentration at these sites (table 3). More recently, during 2000–2010, trends in FN nitrate concentration and flux during the spring are similar to annual trends at most sites (tables 3 and 4). Notable differences include the Iowa River (IOWA-WAP) where, from 2000 through 2010, FN nitrate concentration during the spring changed little (table 4), but decreases during other times of year result in overall decreasing FN nitrate concentration annually (table 3). In the Missouri River (MIZZ-HE), large increases in FN concentration annually (43 percent, table 3) suggest that increases during the spring (26 percent, table 4) are compounded by additional increases during other times of the year.

Site-Specific Observations and Interpretations

At several sites, trends in annual FN nitrate concentration and flux in the 10 years since 2000 differed from trends evident between 1980 and 2000 (table 3). Trends in FN nitrate reversed or stabilized at some locations; FN nitrate trends transitioned from increasing or near-flat to decreasing in the Iowa River (IOWA-WAP) and Illinois River (ILLI-VC). Downstream from these tributaries, FN nitrate trends transitioned from slightly increasing to near flat in the Mississippi River (MSSP-GR). At other sites, FN nitrate began to increase or increased more strongly around 2000. In the lower Mississippi River (MSSP-TH and MSSP-OUT), FN nitrate trends transitioned from near-flat to slightly increasing, and for FN concentration and flux in the Missouri River (MIZZ-HE) and FN flux in the upper Mississippi River (MSSP-CL), increasing FN nitrate trends accelerated. These recent changes in nitrate trends are explored using contour plots that depict the expected value of nitrate concentration as a function of time, season, and streamflow from 2000 through 2010.

Table 3. The 20-year (1980–2000) and 10-year (2000–2010) annual trends in flow-normalized nitrate concentration and flux at eight sites in the Mississippi River Basin.

[mg/L, milligrams per liter; kg/yr, kilograms per year; Increasing and decreasing trends are greater than or equal to ±10 percent, respectively, strongly increasing trends are greater than 40 percent, and trends with minimal change are within ±10 percent]

Site	20-Year annual trends (1980–2000)				10-Year annual trends (2000–2010)			
	FN concentration		FN flux		FN concentration		FN flux	
	Change (mg/L)	Trend (percent change)	Change (10⁶ kg/yr)	Trend (percent change)	Change (mg/L)	Trend (percent change)	Change (10⁶ kg/yr)	Trend (percent change)
MSSP-CL	0.38	Increasing (32)	17	Increasing (23)	0.46	Increasing (29)	23	Increasing (26)
IOWA-WAP	−0.06	Minimal change (−1)	−1	Minimal change (−2)	−0.54	Decreasing (−10)	−8	Decreasing (−13)
ILLI-VC	0.34	Minimal change (9)	14	Increasing (14)	−0.90	Decreasing (−21)	−29	Decreasing (−25)
MSSP-GR	0.27	Increasing (10)	36	Increasing (11)	0.18	Minimal change (6)	3	Minimal change (1)
MIZZ-HE	0.24	Increasing (25)	16	Increasing (17)	0.53	Strongly increasing (43)	26	Increasing (23)
MSSP-TH	0.10	Minimal change (5)	4	Minimal change (1)	0.27	Increasing (13)	36	Minimal change (7)
OHIO-GRCH	0.05	Minimal change (5)	4	Minimal change (1)	−0.07	Minimal change (−6)	−20	Minimal change (−6)
MSSP-OUT	0.05	Minimal change (4)	34	Minimal change (4)	0.16	Increasing (12)	85	Increasing (10)

Table 4. The 20-year (1980–2000) and 10-year (2000–2010) spring trends in flow-normalized nitrate concentration and flux at eight sites in the Mississippi River Basin.

[mg/L, milligrams per liter; kg/yr, kilograms per year; Increasing and decreasing trends are greater than or equal to ±10 percent, respectively, strongly increasing trends are greater than 40 percent, and trends with minimal change are within ±10 percent. Spring trends only consider a 91-day period from April through June]

Site	20-Year spring trends (1980–2000)				10-Year spring trends (2000–2010)			
	FN concentration		FN flux		FN concentration		FN flux	
	Change (mg/L)	Trend (percent change)	Change (10⁶ kg/yr)	Trend (percent change)	Change (mg/L)	Trend (percent change)	Change (10⁶ kg/yr)	Trend (percent change)
MSSP-CL	0.38	Increasing (31)	5	Increasing (16)	0.57	Increasing (35)	12	Increasing (34)
IOWA-WAP	0.67	Increasing (11)	3	Increasing (10)	−0.24	Minimal change (−4)	−3	Decreasing (−10)
ILLI-VC	0.41	Minimal change (8)	5	Increasing (10)	−1.17	Decreasing (−22)	−11	Decreasing (−23)
MSSP-GR	0.47	Increasing (15)	23	Increasing (16)	0.25	Minimal change (7)	4	Minimal change (2)
MIZZ-HE	0.35	Increasing (26)	8	Increasing (19)	0.43	Increasing (26)	6	Increasing (12)
MSSP-TH	0.14	Minimal change (6)	−2	Minimal change (−1)	0.28	Increasing (11)	14	Minimal change (6)
OHIO-GRCH	0.18	Increasing (16)	10	Increasing (11)	−0.12	Minimal change (−9)	−9	Minimal change (−9)
MSSP-OUT	0.15	Increasing (10)	24	Minimal change (7)	0.23	Increasing (14)	30	Minimal change (9)

Mississippi River at Clinton, Iowa

Flow-Normalized Nitrate Concentration and Flux

From 1980 to 2010, the largest percentage increase in FN nitrate flux (55 percent) occurred at Mississippi River at Clinton, Iowa (MSSP-CL), and increases in FN nitrate concentration (70 percent) were comparably large (table 2). FN nitrate concentration was low in 1980 (approximately 1 mg/L), but FN concentrations have generally increased to more than 2 mg/L during the following three decades (fig. 2). FN nitrate concentration increased by approximately 0.5 mg/L between 1980 and 2000 and again between 2000 and 2010, though concentration generally changed little between 1990 and 2000.

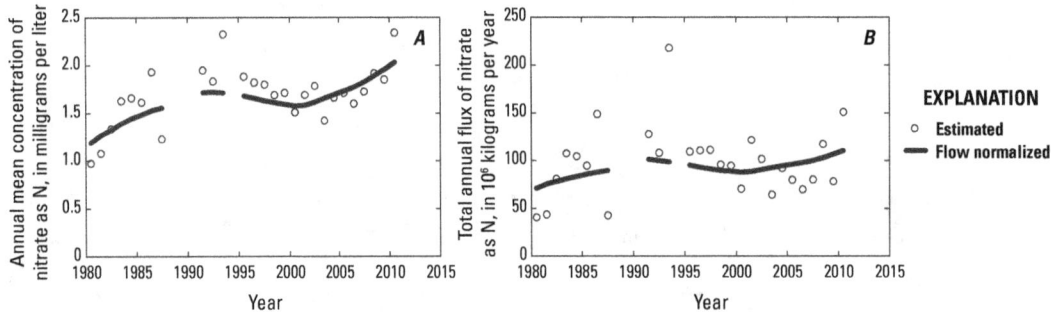

Figure 2.   (*A*) Annual mean estimated concentration (circles) and flow-normalized concentration (solid line) and (*B*) total annual estimated flux (circles) and flow-normalized flux (solid line) from 1980 through 2010 for the Mississippi River at Clinton, Iowa (MSSP-CL).

Comparison of Nitrate Concentrations over Time and with Streamflow

At Mississippi River at Clinton, Iowa (MSSP-CL), between 2000 and 2010, nitrate concentrations increased by approximately 1 mg/L across all streamflows during the winter and spring yet changed little during the summer and fall (fig. 3). Increasing concentration during the winter and spring coincides with an overall increase in annual FN nitrate of about 0.5 mg/L, from 2000 through 2010 (table 3). The highest nitrate concentrations at MSSP-CL consistently occur during the highest streamflows in the winter and appear to increase in intensity over time.

Figure 3.   Expected nitrate concentrations at Mississippi River at Clinton, Iowa (MSSP-CL) from 2000 through 2010. Thin black lines show smoothed estimates of the 5th and 95th percentiles of streamflow. Vertical gray lines indicate January 1 of each year.

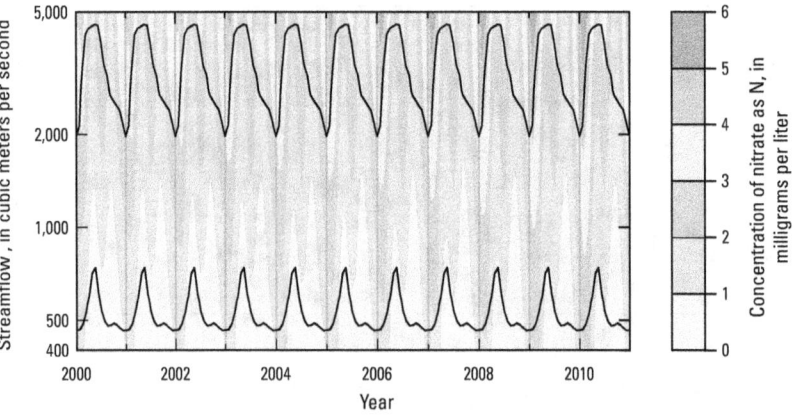

Iowa River at Wapello, Iowa

Flow-Normalized Nitrate Concentration and Flux

FN nitrate concentration and flux decreased (−11 and −15 percent, respectively) from 1980 to 2010 at Iowa River at Wapello, Iowa (IOWA-WAP). FN nitrate was generally stable until early 2000 and then declined slightly through 2010 (fig. 4). In 1980, this site had the highest FN nitrate concentration of any study site (approximately 5 mg/L); even though FN nitrate is decreasing at this site, concentrations still remain higher than at any other site.

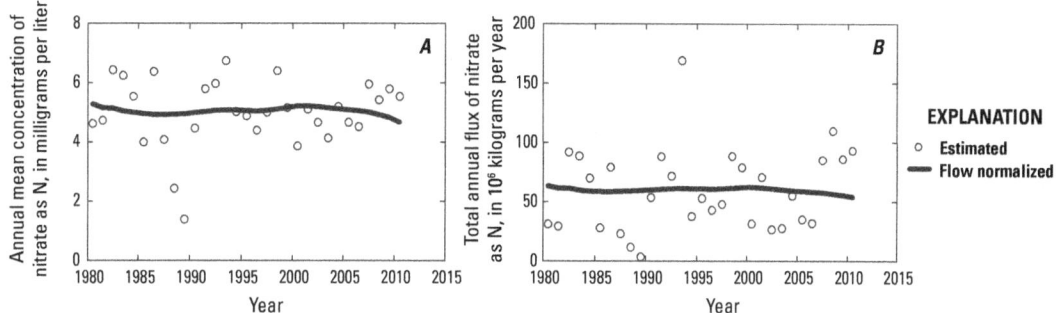

Figure 4. (*A*) Annual mean estimated concentration (circles) and flow-normalized concentration (solid line) and (*B*) total annual estimated flux (circles) and flow-normalized flux (solid line) from 1980 through 2010 for the Iowa River at Wapello, Iowa (IOWA-WAP).

Comparison of Nitrate Concentrations over Time and with Streamflow

From 2000 through 2010, nitrate concentration at Iowa River at Wapello, Iowa (IOWA-WAP) decreased by about 1 to 2 mg/L at moderate and high streamflows across all seasons (fig. 5). Two pulses of elevated nitrate concentrations, centered around the months of May and November of each year, are distinct in the early 2000s but become less prominent over time. This finding suggests a reduction in nitrogen losses from farm fields during high streamflow conditions associated with fall and spring fertilizer application periods and may be related to conservation practices that have been implemented in the State of Iowa during the past decade (Mississippi River/Gulf of Mexico Watershed Nutrient Task Force, 2011). However, increases of about 1 mg/L also occurred at low streamflows during the winter and spring, which, at least during the spring, may partly offset decreases in concentration and result in minimal change of average spring concentration during 2000–2010 (table 4). Increases in nitrate concentration during low flows at this site and others may be because of legacy nitrate from groundwater, which could continue to influence base-flow concentrations for decades (Tesoriero and others, 2013).

Figure 5. Expected nitrate concentrations at Iowa River at Wapello, Iowa (IOWA-WAP) from 2000 through 2010. Thin black lines show smoothed estimates of the 5th and 95th percentiles of streamflow. Vertical gray lines indicate January 1 of each year.

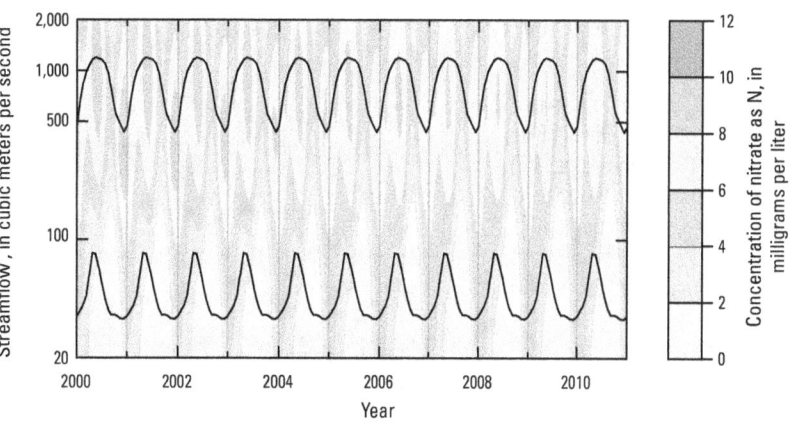

Illinois River at Valley City, Illinois

Flow-Normalized Nitrate Concentration and Flux

FN nitrate concentration and flux decreased (–14 percent) at Illinois River at Valley City, Illinois (ILLI-VC) from 1980 to 2010 (table 2). FN nitrate was relatively stable from 1980 through 2000, and considerable decreases in FN nitrate were observed from 2000 through 2010 (–21 percent for concentration and –25 percent for flux; fig. 6). FN nitrate concentration was high in 1980 (approximately 4 mg/L), second only to the concentration in the Iowa River (IOWA-WAP).

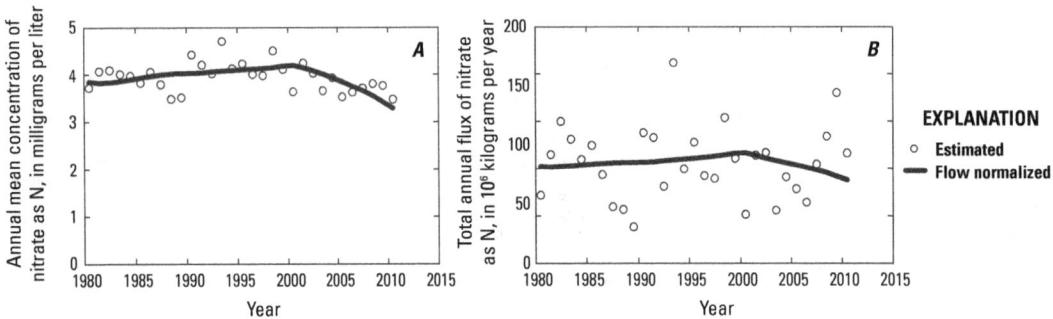

Figure 6. (*A*) Annual mean estimated concentration (circles) and flow-normalized concentration (solid line) and (*B*) total annual estimated flux (circles) and flow-normalized flux (solid line) from 1980 through 2010 for the Illinois River at Valley City, Illinois (ILLI-VC).

Comparison of Nitrate Concentrations over Time and with Streamflow

Nitrate concentration decreased considerably across all streamflows from 2000 through 2010 at Illinois River at Valley City, Illinois (ILLI-VC). The greatest decreases in nitrate, as much as 2 mg/L, occurred during high streamflows across the entire year, and decreases during moderate and low streamflows were also substantial (approximately 0.5 mg/L; fig. 7). The highest nitrate concentrations continue to occur during the winter and spring under high streamflow conditions, though they are becoming less severe over time. Declining nitrate concentrations during high flows may indicate that conservation efforts to reduce nitrogen losses in runoff from farm fields are achieving results (Mississippi River/Gulf of Mexico Watershed Nutrient Task Force, 2011; U.S. Department of Agriculture, Natural Resources Conservation Service, 2012). Additionally, reductions at low flows in the Illinois River suggest legacy nitrate may be less influential in this subbasin as compared to the Iowa River (IOWA-WAP) or that point source discharges to the Illinois River have changed over the past decade.

Figure 7. Expected nitrate concentrations at Illinois River at Valley City, Illinois (ILLI-VC) from 2000 through 2010. Thin black lines show smoothed estimates of the 5th and 95th percentiles of streamflow. Vertical gray lines indicate January 1 of each year.

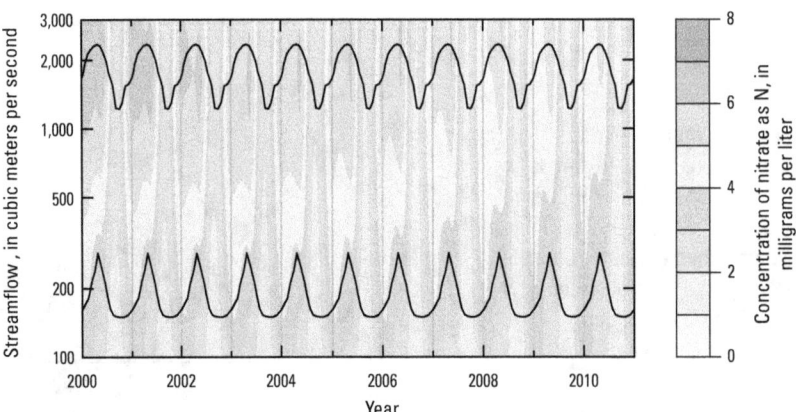

Mississippi River below Grafton, Illinois

Flow-Normalized Nitrate Concentration and Flux

FN nitrate concentration and flux increased slightly from 1980 to 2010 (17 and 11 percent, respectively) at Mississippi River below Grafton, Illinois (MSSP-GR) (table 2). This site has the highest FN nitrate concentration of any of the study sites located on the Mississippi River (approximately 2.5 mg/L in 1980 and approximately 3 mg/L in 2010; fig. 8), reflective of upstream contributions from intensely farmed subbasins with high nitrate concentrations, such as the subbasins of the Iowa (IOWA-WAP) and Illinois (ILLI-VC) Rivers. Nitrate changed little at MSSP-GR from 2000 onward, possibly reflecting the integration of increasing nitrate from some portions of the basin, such as the upper Mississippi River (MSSP-CL) and the intervening subbasin above MSSP-GR, and decreasing nitrate in others, such as the Iowa and Illinois River Basins.

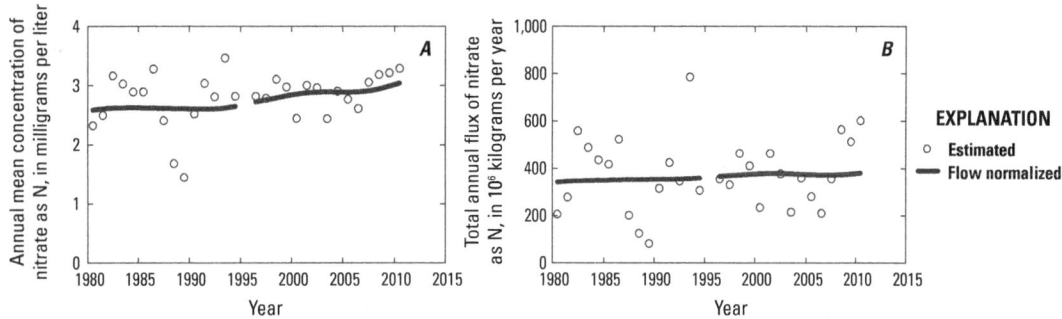

Figure 8.   (*A*) Annual mean estimated concentration (circles) and flow-normalized concentration (solid line) and (*B*) total annual estimated flux (circles) and flow-normalized flux (solid line) from 1980 through 2010 for the Mississippi River below Grafton, Illinois (MSSP-GR).

Comparison of Nitrate Concentrations over Time and with Streamflow

Mississippi River below Grafton, Illinois (MSSP-GR) had minimal change in nitrate concentration from 2000 through 2010 (fig. 9). Each year, two pulses of elevated nitrate concentration at moderate and high streamflows, centered around the months of January and May, continued to occur. There is evidence for a slight decrease in nitrate during high streamflows, particularly during the summer and fall, but this decrease is offset by increases at moderate and low streamflows primarily during the winter and spring (fig. 9).

Figure 9.   Expected nitrate concentrations at Mississippi River below Grafton, Illinois (MSSP-GR) from 2000 through 2010. Thin black lines show smoothed estimates of the 5th and 95th percentiles of streamflow. Vertical gray lines indicate January 1 of each year.

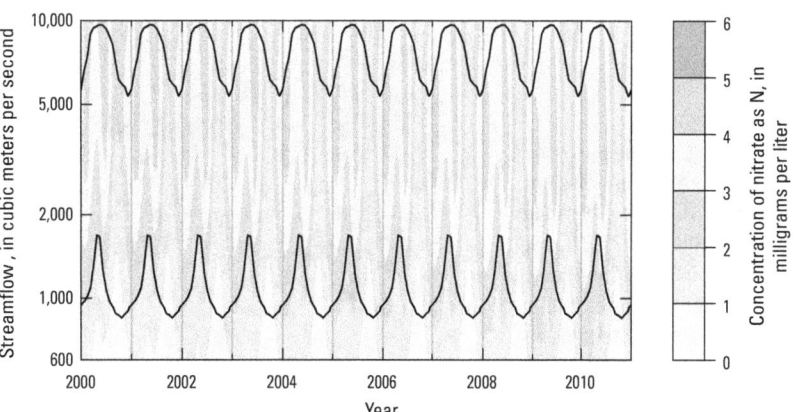

Missouri River at Hermann, Missouri

Flow-Normalized Nitrate Concentration and Flux

From 1980 through 2010, the largest percentage increase in FN nitrate concentration at any of the study sites occurred at Missouri River at Hermann, Missouri (MIZZ-HE) (79 percent; table 2). FN flux also increased substantially at this site over the same period (45 percent). FN nitrate concentration at this site in 1980 was approximately 1 mg/L, among the lowest of any site in this study. FN nitrate concentration and flux increased consistently from 1980 through 2010, though nitrate increased at a faster rate from the early 2000s onward (fig. 10).

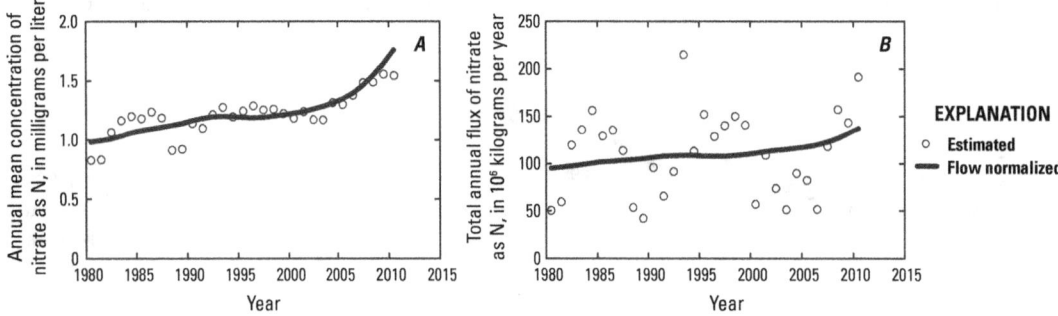

Figure 10.   (*A*) Annual mean estimated concentration (circles) and flow-normalized concentration (solid line) and (*B*) total annual estimated flux (circles) and flow-normalized flux (solid line) from 1980 through 2010 for the Missouri River at Hermann, Missouri (MIZZ-HE).

Comparison of Nitrate Concentrations over Time and with Streamflow

Since 2000, nitrate concentration increased primarily during moderate and low streamflows at Missouri River at Hermann, Missouri (MIZZ-HE). This is particularly evident during spring and summer low streamflows (for example, at discharge values around 1,000 cubic meters per second [m³/s]) when nitrate concentration increased from just under 2 mg/L in the early 2000s to greater than 3.5 mg/L in 2010 (fig. 11). These increases at low streamflows may be due to legacy nitrate from groundwater (Tesoriero and others, 2013) or could be related to increases from point source discharges in the basin. Slight increases in nitrate also occurred across all streamflows during the winter, though there was little change in the nitrate concentration at high discharges (about 5,000 m³/s) during the rest of the year.

Figure 11.   Expected nitrate concentrations at Missouri River at Hermann, Missouri (MIZZ-HE) from 2000 through 2010. Thin black lines show smoothed estimates of the 5th and 95th percentiles of streamflow. Vertical gray lines indicate January 1 of each year.

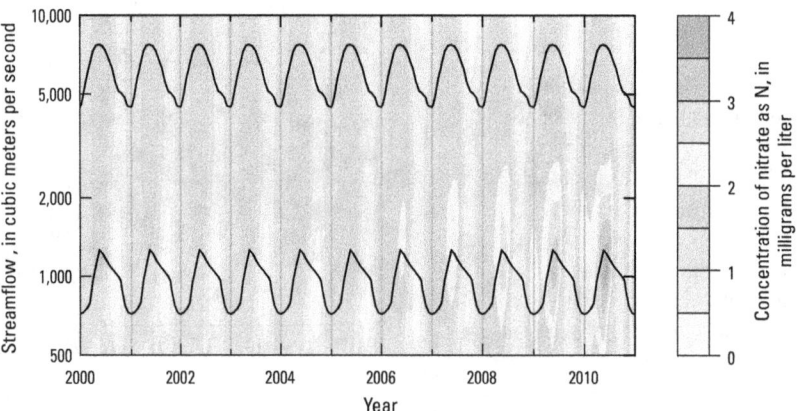

Mississippi River at Thebes, Illinois

Flow-Normalized Nitrate Concentration and Flux

From 1980 through 2010, FN nitrate concentration increased (19 percent), and FN flux experienced minimal change (8 percent) at Mississippi River at Thebes (MSSP-TH; table 2; fig. 12). Most of the increase in FN concentration occurred recently (2000–2010; 0.27 mg/L), as compared to the previous 20 years when concentration changed little (0.10 mg/L).

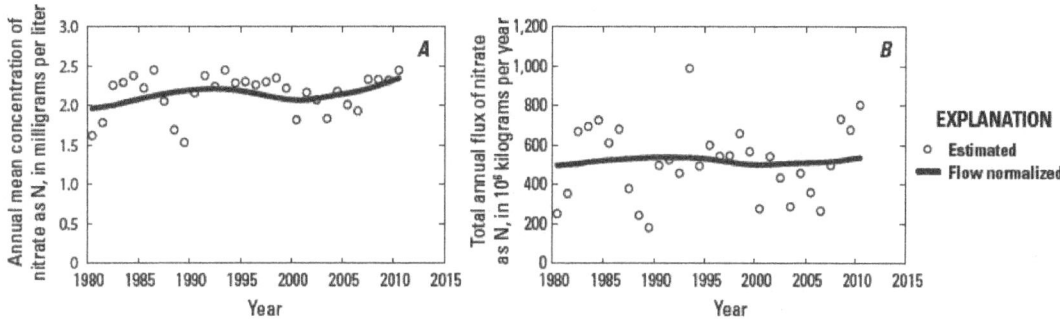

Figure 12 An a e m t d o n c le l w o a concentration (sol d line) and (*B*) total annual estimated flux (circles) and flow-normalized flux (solid line) from 1980 hrough 2010 for the Mississipp River at Thebes, I linois (MSSP-TH).

Comparison of Nitrate Concentrations over Time and with Streamflow

Nitrate concentration increased across most streamflows from 2000 through 2010 at Mississippi River at Thebes, Illinois (MSSP-TH). These increases are most prominent during low and moderate streamflows in the fall, winter, and spring (fig. 13) and may be related to changes in nitrate from groundwater or point sources. Some of these increases in nitrate are offset by slight decreases at moderate and high streamflows during the summer. These increases and decreases of nitrate concentration by season and streamflow at MSSP-TH reflect a mixture of the different trends observed at upstream sites, such as decreases in nitrate during high streamflows in the Mississippi River (MSSP-GR) and increases in nitrate during low streamflow in the Missouri River (MIZZ-HE) and MSSP-GR.

Figure 13. Expected nitrate concentrations at Mississippi River at Thebes, Illinois (MSSP-TH) from 2000 through 2010. Thin black lines show smoothed estimates of the 5th and 95th percentiles of streamflow. Vertical gray lines indicate January 1 of each year.

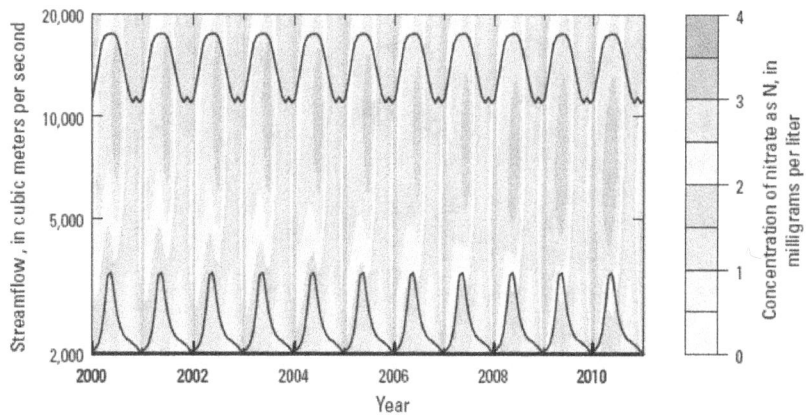

Ohio River at Dam 53 near Grand Chain, Illinois

Flow-Normalized Nitrate Concentration and Flux

FN nitrate concentration and flux have changed little during 1980–2010 at Ohio River at Dam 53, near Grand Chain, Illinois (OHIO-GRCH) (–2 and –5 percent, respectively). FN nitrate concentration is stable over much of the period of record, though a slight decrease may have occurred since the late 2000s (fig. 14). FN nitrate concentration was low in 1980 (approximately 1 mg/L) and remained low during the following three decades.

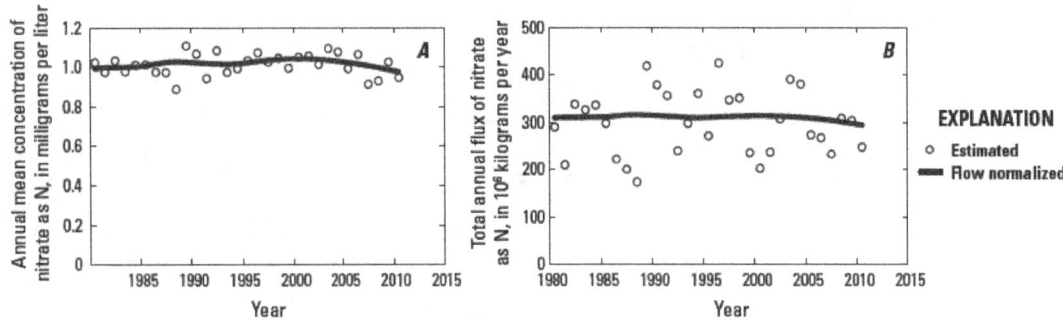

Figure 14.   (*A*) Annual mean estimated concentration (circles) and flow-normalized concentration (solid line) and (*B*) total annual estimated flux (circles) and flow-normalized flux (solid line) from 1980 through 2010 for the Ohio River at Dam 53 near Grand Chain, Illinois (OHIO-GRCH)

Comparison of Nitrate Concentrations over Time and with Streamflow

At Ohio River at Dam 53 near Grand Chain, Illinois (OHIO-GRCH), nitrate concentrations were the lowest observed at any site (fig. 15). Slight decreases of less than 0.25 mg/L generally occurred from 2000 through 2010, across all streamflows. The highest nitrate concentrations (less than 2 mg/L) were primarily at moderate streamflows during one of two concentration pulses, centered around the months of January and May, and over time appear to have lessened in intensity. However, because nitrate concentrations at OHIO-GRCH were initially low, changes over time at particular streamflows are difficult to discern.

Figure 15.   Expected nitrate concentrations at Ohio River at Dam 53 near Grand Chain, Illinois (OHIO-GRCH) from 2000 through 2010. Thin black lines show smoothed estimates of the 5th and 95th percentiles of streamflow. Vertical gray lines indicate January 1 of each year.

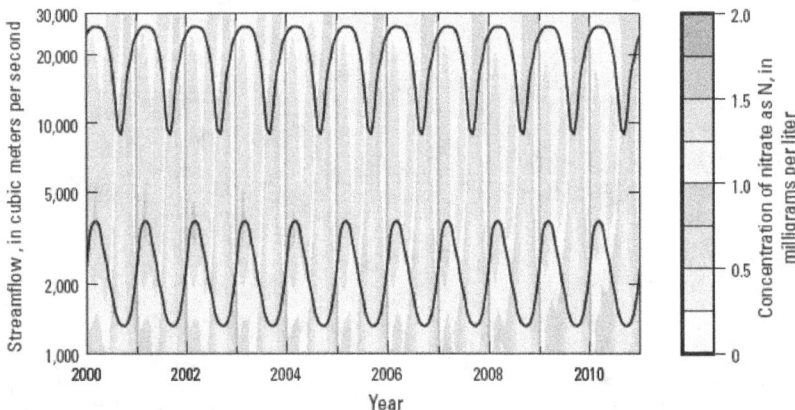

Mississippi River above Old River Outflow Channel, Louisiana

Flow-Normalized Nitrate Concentration and Flux

FN nitrate concentration and flux have increased slightly during 1980–2010 (17 and 14 percent, respectively; table 2) at Mississippi River above Old River Outflow Channel, Louisiana (MSSP-OUT). During this time, annual mean FN nitrate concentrations ranged from 1.2 to 1.5 mg/L. During 1980–2000, nitrate changed little, whereas from 2000 onward, FN nitrate concentration and flux increased slightly by 12 and 10 percent, respectively (fig. 16). The recent increase in FN nitrate concentration and flux at MSSP-OUT indicates that even though decreases in FN nitrate concentration have been observed at smaller subbasins with historically elevated nitrate (IOWA-WAP and ILLI-VC), these decreases appear to be offset by increases in nitrate from other areas of the basin, such as the upper Mississippi River (MSSP-CL), the Missouri River (MIZZ-HE), and the intervening basin above MSSP-OUT. The recent upturn in FN nitrate concentration and flux at MSSP-OUT may not develop into a sustained increasing trend; however, the upturn does suggest that conditions within the MRB are likely changing and how this trend develops in the coming years will have important management implications.

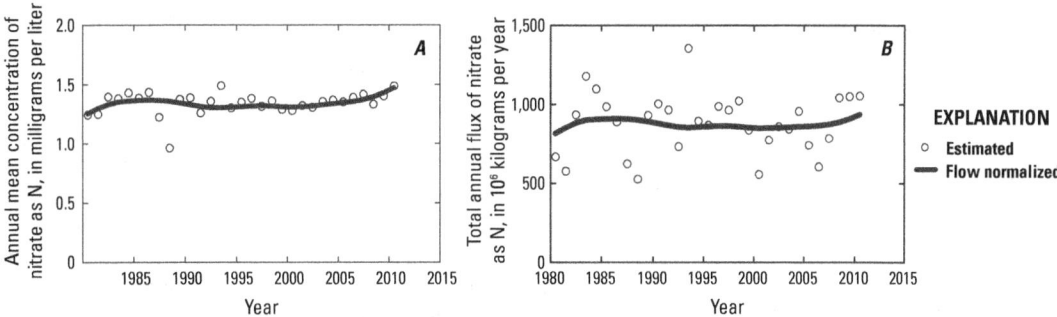

Figure 16. (*A*) Annual mean estimated concentration (circles) and flow-normalized concentration (solid line) and (*B*) total annual estimated flux (circles) and flow-normalized flux (solid line) from 1980 through 2010 for the Mississippi River above Old River Outflow Channel, Louisiana (MSSP OUT).

Comparison of Nitrate Concentrations over Time and with Streamflow

Since 2000, nitrate concentration has increased slightly across a range of streamflows and seasons at Mississippi River above Old River Outflow Channel, Louisiana (MSSP-OUT). During the fall and winter, nitrate increased slightly (less than 0.5 mg/L) across all streamflows (fig. 17). Nitrate also increased during the spring, by about 1 mg/L, at low streamflows. These increases in nitrate appear to be partly offset by slight decreases (less than 0.5 mg/L) at moderate and high streamflows in the summer. The highest nitrate concentrations typically occur during low and moderate streamflows in April and May. Increases in nitrate concentration during low streamflows at this site and at many other sites in the MRB (except ILLI-VC and OHIO-GRCH) suggest that contributions from point sources, such as wastewater treatment plants, or legacy nitrate from groundwater are becoming important influences throughout the basin as a whole, even as concentrations have decreased in intensely farmed subbasins, such as the Iowa (IOWA-WAP) and Illinois (ILLI-VC) Rivers, which historically have been important contributors of nitrogen (Goolsby and Battaglin, 2001).

Figure 17. Expected nitrate concentrations at Old River Outflow Channel, Louisiana (MSSP-OUT) from 2000 through 2010. Thin black lines show smoothed estimates of the 5th and 95th percentiles of streamflow. Vertical gray lines indicate January 1 of each year.

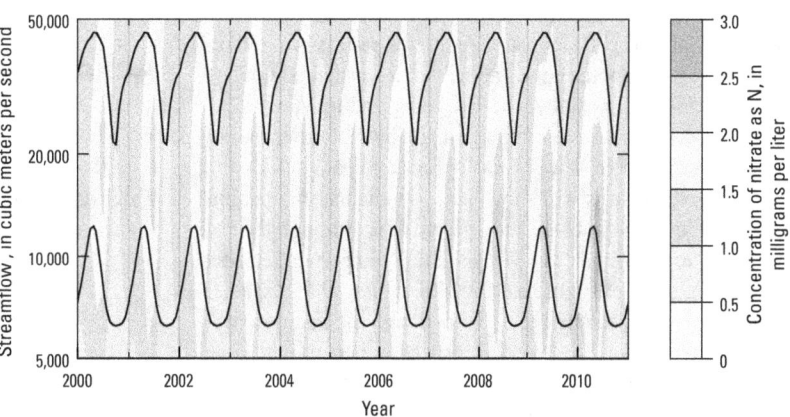

Effect of Calibration Period on Estimates of Nitrate Concentration and Flux

The 1980 through 2008 estimates of nitrate concentration and flux presented in this report differ slightly from those published by Sprague and others (2011) for these eight sites. Although this difference partly reflects minor modifications of the WRTDS method (discussed in the Methods section), a more important factor is the addition of 2 years of streamflow and nitrate concentration to the calibration datasets. Any smoothing approach, such as WRTDS, tends to produce less stable estimates at the end of the period of record (Hirsch and others, 2010). With the addition of 2 years of data, the estimates of concentration and flux that were previously at the end of the record are now informed by later as well as earlier observations. If the behavior of the system is changing rapidly, new estimates for the final few years of the original period of record can change by several percent. Estimates near the middle of the record are generally stable and change very little (often less than 1 percent) as more data are added (Hirsch and others, 2010). The addition of new streamflow data may also result in changes in FN values because of the changing characterization of the probability distribution of streamflow. Thus, the addition of more streamflow data can result in modest changes in FN values across the period of record but have little influence on the slopes of estimated trends over the earlier years of the record.

Annual estimates of nitrate published by Sprague and others (2011) and this report were compared at all sites and for all possible years (1980–2008), and in general the differences between values were small. For annual concentration and flux (non-FN), the average of the absolute percent differences between estimates reported in Sprague and others (2011) and this report were approximately 1 percent. For annual FN concentration and flux, the average absolute percent differences between estimates were slightly larger, at 1.5 percent and 2.8 percent, respectively. The largest difference in FN flux over all of the 227 values that could be compared (8 sites times 29 years minus 5 missing years) is a difference of 10.1 percent at the site on the Missouri River (MIZZ-HE) where the 2008 estimate decreased from 141×10^6 kg/yr in the prior study to 127×10^6 kg/yr in this study. In this case, the prior study (Sprague and others, 2011) overstated the rate of change around 2008 as compared to the results presented in this report, but both reports agree that a substantial upward trend was taking place over the few years leading up to 2008. This case demonstrates an important attribute of the smoothing approach used in WRTDS (as opposed to parametric approach); the results will show incipient changes happening in the last few years of the study period and, although the rates of change for these last few years are more uncertain for sites undergoing rapid changes as compared to sites with stable trends, they will become better defined with the addition of 1 or 2 years of data. Parametric approaches, because they are typically constrained to follow a linear or quadratic time trend, will not provide any indication of changes as they develop.

Summary and Conclusions

Updated estimates of flow-normalized nitrate concentration and flux indicate decreasing trends in the Iowa River (–11 and –15 percent, respectively) and Illinois River (–14 percent, each) during 1980–2010. These decreasing trends are a recent development (2000–2010) and became readily apparent when nitrate concentration and flux estimates were extended from 2008 through 2010, suggesting that change within these basins has been occurring over the past 10 years. In the Ohio River, flow-normalized nitrate concentration and flux have been relatively low and stable, changing less than 10 percent, from 1980 through 2010. Flow-normalized nitrate concentration at the remaining five sites increased from 1980 through 2010. In the upper Mississippi River and Missouri River, flow-normalized nitrate concentration and flux have increased steadily during 1980–2000 and, in both cases, have accelerated since 2000. At the two most downstream sites on the Mississippi River, increasing trends in flow-normalized nitrate concentration are a relatively new development and at least partly reflect increases occurring in the Missouri River and the upper Mississippi River, in addition to other tributaries not investigated in this study. The recent upturn in nitrate at the outlet of the Mississippi River has the potential to develop into a sustained increase, which is particularly important because it suggests that despite declining nitrate concentrations in smaller tributaries that historically have had high nitrate concentrations, such as in the Iowa and Illinois Rivers, nitrate appears to be increasing within the Mississippi River Basin as a whole. Increases in nitrate concentration during low flows were evident at the outflow of the Mississippi River, which reflect similar increases during low flows observed at many study sites in the MRB (the Illinois and Ohio Rivers excepted). Basin-wide increases in nitrate during low flows may have important management implications. If contributions are predominantly from point sources, the potential to affect concentrations is great, by means of upgrades to wastewater treatment systems or stricter regulation of point sources; however, if legacy nitrate from groundwater is the predominate source, then it may take decades before substantial decreases in concentrations occur, regardless of improvements in agricultural practices.

Acknowledgments

The authors thank the numerous U.S. Geological Survey (USGS) field personnel who collected the decades of surface-water data used in this report, and Brent Aulenbach (USGS) for his assistance in compiling and preparing the data.

References Cited

Aulenbach, B.T., Buxton, H.T., Battaglin, W.A., and Coupe, R.H., 2007, Flow and nutrient fluxes of the Mississippi-Atchafalaya River Basin and subbasins for the period of record through 2005: U.S. Geological Survey Open-File Report 2007–1080.

Duan, N., 1983, Smearing estimate—A nonparameteric retransformation method: Journal of the American Statistical Association, v. 78, p. 605–610.

Garrett, J.D., 2012, Concentrations, loads, and yields of selected constituents from major tributaries of the Mississippi and Missouri Rivers in Iowa, water years 2004–2008: U.S. Geological Survey Scientific Investigations Report 2012–5240, 61 p. (Also available online at *http://pubs.usgs.gov/sir/2012/5240/*).

Goolsby, D.A., and Battaglin, W.A., 2001, Long-term changes in concentrations and flux of nitrogen in the Mississippi River Basin, USA: Hydrological Processes, v. 15, p. 1209–1226.

Goolsby, D.A., Battaglin, W.A., Aulenbach, B.T., and Hooper, R.P., 2000, Nitrogen flux and sources in the Mississippi River Basin: The Science of the Total Environment, v. 248, p. 75–86.

Hirsch, R.M., Moyer, D.L., and Archfield, S.A., 2010, Weighted regressions on time, discharge, and season (WRTDS), with an application to Chesapeake Bay River inputs: Journal of the American Water Resources Association, v. 46, no. 5, p. 857–880.

McIssac, G.F., David, M.B., Gertner, G.Z., and Goolsby, D.A., 2001, Nitrate flux in the Mississippi River: Nature, v. 414, p. 166–167.

Mississippi River/Gulf of Mexico Watershed Nutrient Task Force, 2008, Gulf hypoxia action plan 2008 for reducing, mitigating, and controlling hypoxia in the northern Gulf of Mexico and improving water quality in the Mississippi River Basin: Washington, D.C., U.S. Environmental Protection Agency.

Mississippi River/Gulf of Mexico Watershed Nutrient Task Force, 2011, Moving forward on Gulf hypoxia—Annual report 2011: Washington, D.C.

Moyer, D.L., Hirsch, R.M., and Hyer, K.E., 2012, Comparison of two regression-based approaches for determining nutrient and sediment fluxes and trends in the Chesapeake Bay watershed: U.S. Geological Survey Scientific Investigations Report 2012–5244, 118 p. (Also available online at *http://pubs.usgs.gov/sir/2012/5244.*)

Rabalais, N.N., Turner, R.E., and Wiseman, W.J., 2002, Gulf of Mexico hypoxia, a k.a. "The Dead Zone": Annual Review of Ecology and Systematics, v. 33, p. 235–263.

Scavia, D., Rabalais, N.N., Turner, R.E., Justic, D., and Wiseman, W.J., 2003, Predicting the response of Gulf of Mexico hypoxia to variations in Mississippi River nitrogen load: Limnology and Oceanography, v. 48, no. 3, p. 951–956.

Sprague, L.A., Hirsch, R.M., and Aulenbach, B.T., 2011, Nitrate in the Mississippi River and its tributaries, 1980 to 2008—Are we making progress?: Environmental Science & Technology, v. 45, p. 7209–7216.

Stenback, G.A., Crumpton, W.G., Schilling, K.E., and Helmers, M.J., 2011, Rating curve estimation of nutrient loads in Iowa rivers: Journal of Hydrology, v. 396, p. 158–169.

Tesoriero, A.J., Duff, J.H., Saad, D.A., Spahr, N.E., and Wolock, D.M., 2013, Vulnerability of streams to legacy nitrate sources. Environmental Science & Technology, v. 47, p. 3623–3629.

Tukey, J.W., 1977, Exploratory data analysis: Reading, Massachusetts, Addison-Wesley.

Turner, R.E., Rabalais, N.N., and Justic, D., 2006, Predicting summer hypoxia in the northern Gulf of Mexico—Riverine N, P, and Si loading: Marine Pollution Bulletin, v. 52, p. 139–148.

Turner, R.E., Rabalais, N.N., and Justic, D., 2012, Predicting summer hypoxia in the northern Gulf of Mexico—Redux: Marine Pollution Bulletin, v. 64, p. 319–324.

U.S. Department of Agriculture, Natural Resources Conservation Service, 2012, Assessment of the effects of conservation practices on cultivated cropland in the upper Mississippi River Basin, p. 187.

Appendix 1. Annual mean estimated and flow-normalized nitrate concentration and total annual estimated and flow-normalized nitrate flux from WRTDS for 1980 through 2010, by calendar years for eight sites in the Mississippi River Basin.

[WRTDS, weighted regressions on time, discharge, and season model; —, estimates not reported because too few samples were collected in that year; km², square kilometers; mg/L, milligrams per liter; kg/yr, kilograms per year]

Site	Drainage area, in km²	Calendar year	Annual mean estimated concentration, in mg/L	Annual mean flow-normalized concentration, in mg/L	Total annual estimated flux, in 10⁶ kg/yr	Total annual flow-normalized flux, in 10⁶ kg/yr
MSSP-CL	221,703	1980	0.98	1.19	40.6	71.1
MSSP-CL	221,703	1981	1.08	1.27	43.6	75.7
MSSP-CL	221,703	1982	1.34	1.33	81.1	78.3
MSSP-CL	221,703	1983	1.63	1.40	107.6	81.1
MSSP-CL	221,703	1984	1.66	1.44	104.2	83.2
MSSP-CL	221,703	1985	1.62	1.49	94.5	85.7
MSSP-CL	221,703	1986	1.93	1.53	148.8	87.6
MSSP-CL	221,703	1987	1.24	1.56	42.8	89.3
MSSP-CL	221,703	1988	—	—	—	—
MSSP-CL	221,703	1989	—	—	—	—
MSSP-CL	221,703	1990	—	—	—	—
MSSP-CL	221,703	1991	1.95	1.72	127.9	101.0
MSSP-CL	221,703	1992	1.83	1.72	107.8	99.8
MSSP-CL	221,703	1993	2.33	1.71	218.4	98.6
MSSP-CL	221,703	1994	—	—	—	—
MSSP-CL	221,703	1995	1.88	1.68	109.1	95.1
MSSP-CL	221,703	1996	1.82	1.66	110.1	92.9
MSSP-CL	221,703	1997	1.80	1.63	110.8	90.9
MSSP-CL	221,703	1998	1.69	1.61	95.6	89.6
MSSP-CL	221,703	1999	1.71	1.59	94.5	88.6
MSSP-CL	221,703	2000	1.51	1.58	70.3	87.8
MSSP-CL	221,703	2001	1.69	1.58	121.6	88.5
MSSP-CL	221,703	2002	1.78	1.61	101.4	90.5
MSSP-CL	221,703	2003	1.42	1.66	64.2	92.6
MSSP-CL	221,703	2004	1.66	1.69	91.9	94.2
MSSP-CL	221,703	2005	1.71	1.73	79.8	95.9
MSSP-CL	221,703	2006	1.60	1.77	69.9	97.5
MSSP-CL	221,703	2007	1.73	1.83	80.1	100.0
MSSP-CL	221,703	2008	1.92	1.89	117.4	103.2
MSSP-CL	221,703	2009	1.85	1.96	78.2	106.7
MSSP-CL	221,703	2010	2.34	2.03	150.9	110.4
IOWA-WAP	32,375	1980	4.61	5.28	31.1	63.3
IOWA-WAP	32,375	1981	4.73	5.16	29.3	61.2
IOWA-WAP	32,375	1982	6.43	5.15	91.9	61.1
IOWA-WAP	32,375	1983	6.26	5.05	88.8	59.6
IOWA-WAP	32,375	1984	5.54	5.01	69.7	58.9

Appendix 1.   Annual mean estimated and flow-normalized nitrate concentration and total annual estimated and flow-normalized nitrate flux from WRTDS for 1980 through 2010, by calendar years for eight sites in the Mississippi River Basin.—Continued

[WRTDS, weighted regressions on time, discharge, and season model; —, estimates not reported because too few samples were collected in that year; km², square kilometers; mg/L, milligrams per liter; kg/yr, kilograms per year]

Site	Drainage area, in km²	Calendar year	Annual mean estimated concentration, in mg/L	Annual mean flow-normalized concentration, in mg/L	Total annual estimated flux, in 10⁶ kg/yr	Total annual flow-normalized flux, in 10⁶ kg/yr
IOWA-WAP	32,375	1985	4.00	4.96	28.0	58.4
IOWA-WAP	32,375	1986	6.38	4.93	78.9	58.1
IOWA-WAP	32,375	1987	4.09	4.93	23.1	58.4
IOWA-WAP	32,375	1988	2.43	4.93	11.7	58.7
IOWA-WAP	32,375	1989	1.39	4.96	3.6	59.3
IOWA-WAP	32,375	1990	4.47	4.99	53.5	59.9
IOWA-WAP	32,375	1991	5.80	5.03	88.0	60.4
IOWA-WAP	32,375	1992	5.97	5.06	71.4	60.7
IOWA-WAP	32,375	1993	6.75	5.08	169.3	60.9
IOWA-WAP	32,375	1994	5.02	5.07	37.5	60.8
IOWA-WAP	32,375	1995	4.88	5.06	52.8	60.5
IOWA-WAP	32,375	1996	4.40	5.04	42.8	60.4
IOWA-WAP	32,375	1997	5.01	5.06	47.8	60.6
IOWA-WAP	32,375	1998	6.40	5.11	88.3	61.2
IOWA-WAP	32,375	1999	5.16	5.17	78.4	61.7
IOWA-WAP	32,375	2000	3.86	5.22	31.4	61.9
IOWA-WAP	32,375	2001	5.11	5.22	70.4	61.6
IOWA-WAP	32,375	2002	4.67	5.20	26.6	60.8
IOWA-WAP	32,375	2003	4.14	5.15	27.6	59.9
IOWA-WAP	32,375	2004	5.19	5.12	54.9	59.2
IOWA-WAP	32,375	2005	4.68	5.09	35.0	58.6
IOWA-WAP	32,375	2006	4.52	5.06	31.8	58.0
IOWA-WAP	32,375	2007	5.95	5.01	84.7	57.3
IOWA-WAP	32,375	2008	5.43	4.93	109.9	56.4
IOWA-WAP	32,375	2009	5.80	4.83	85.7	55.3
IOWA-WAP	32,375	2010	5.54	4.68	93.0	53.9
ILLI-VC	69,264	1980	3.72	3.85	71.9	102.4
ILLI-VC	69,264	1981	4.07	3.83	115.0	102.0
ILLI-VC	69,264	1982	4.09	3.83	150.0	102.3
ILLI-VC	69,264	1983	4.01	3.86	131.2	103.1
ILLI-VC	69,264	1984	3.97	3.89	109.6	103.9
ILLI-VC	69,264	1985	3.83	3.93	124.8	104.8
ILLI-VC	69,264	1986	4.06	3.97	93.3	105.5
ILLI-VC	69,264	1987	3.80	4.00	59.7	106.1
ILLI-VC	69,264	1988	3.48	4.02	56.8	106.4
ILLI-VC	69,264	1989	3.52	4.02	38.3	106.4

Appendix 1. Annual mean estimated and flow-normalized nitrate concentration and total annual estimated and flow-normalized nitrate flux from WRTDS for 1980 through 2010, by calendar years for eight sites in the Mississippi River Basin.—Continued

[WRTDS, weighted regressions on time, discharge, and season model; —, estimates not reported because too few samples were collected in that year; km², square kilometers; mg/L, milligrams per liter; kg/yr, kilograms per year]

Site	Drainage area, in km²	Calendar year	Annual mean estimated concentration, in mg/L	Annual mean flow-normalized concentration, in mg/L	Total annual estimated flux, in 10⁶ kg/yr	Total annual flow-normalized flux, in 10⁶ kg/yr
ILLI-VC	69,264	1990	4.42	4.03	138.0	106.7
ILLI-VC	69,264	1991	4.20	4.04	133.0	107.3
ILLI-VC	69,264	1992	4.02	4.06	80.9	108.2
ILLI-VC	69,264	1993	4.70	4.07	211.8	109.1
ILLI-VC	69,264	1994	4.13	4.09	99.6	110.1
ILLI-VC	69,264	1995	4.23	4.11	128.1	111.1
ILLI-VC	69,264	1996	4.00	4.12	92.6	112.2
ILLI-VC	69,264	1997	3.98	4.12	89.5	113.1
ILLI-VC	69,264	1998	4.50	4.15	153.8	114.7
ILLI-VC	69,264	1999	4.10	4.18	110.8	116.3
ILLI-VC	69,264	2000	3.64	4.19	51.7	116.6
ILLI-VC	69,264	2001	4.24	4.15	114.3	114.2
ILLI-VC	69,264	2002	4.03	4.08	116.9	110.9
ILLI-VC	69,264	2003	3.66	4.00	56.0	108.2
ILLI-VC	69,264	2004	3.93	3.93	91.1	106.2
ILLI-VC	69,264	2005	3.53	3.85	78.5	103.9
ILLI-VC	69,264	2006	3.63	3.76	64.4	101.5
ILLI-VC	69,264	2007	3.71	3.66	104.7	98.7
ILLI-VC	69,264	2008	3.81	3.55	134.2	95.4
ILLI-VC	69,264	2009	3.77	3.42	180.1	91.6
ILLI-VC	69,264	2010	3.48	3.29	116.4	87.7
MSSP-GR	443,665	1980	2.32	2.59	205.4	342.0
MSSP-GR	443,665	1981	2.50	2.61	277.5	345.3
MSSP-GR	443,665	1982	3.17	2.62	559.2	347.3
MSSP-GR	443,665	1983	3.03	2.63	489.3	349.0
MSSP-GR	443,665	1984	2.89	2.63	436.3	349.7
MSSP-GR	443,665	1985	2.89	2.62	418.5	349.9
MSSP-GR	443,665	1986	3.28	2.62	524.1	351.1
MSSP-GR	443,665	1987	2.41	2.62	201.6	351.8
MSSP-GR	443,665	1988	1.68	2.61	125.6	352.8
MSSP-GR	443,665	1989	1.45	2.61	82.4	354.1
MSSP-GR	443,665	1990	2.52	2.60	314.6	354.3
MSSP-GR	443,665	1991	3.04	2.60	425.5	354.5
MSSP-GR	443,665	1992	2.81	2.61	345.2	354.9
MSSP-GR	443,665	1993	3.47	2.62	786.0	356.0
MSSP-GR	443,665	1994	2.82	2.65	307.0	359.2

Appendix 1.   Annual mean estimated and flow-normalized nitrate concentration and total annual estimated and flow-normalized nitrate flux from WRTDS for 1980 through 2010, by calendar years for eight sites in the Mississippi River Basin.—Continued

[WRTDS, weighted regressions on time, discharge, and season model; —, estimates not reported because too few samples were collected in that year; km², square kilometers; mg/L, milligrams per liter; kg/yr, kilograms per year]

Site	Drainage area, in km²	Calendar year	Annual mean estimated concentration, in mg/L	Annual mean flow-normalized concentration, in mg/L	Total annual estimated flux, in 10⁶ kg/yr	Total annual flow-normalized flux, in 10⁶ kg/yr
MSSP-GR	443,665	1995	—	—	—	—
MSSP-GR	443,665	1996	2.82	2.72	355.2	367.5
MSSP-GR	443,665	1997	2.78	2.76	330.1	370.2
MSSP-GR	443,665	1998	3.10	2.79	463.9	372.8
MSSP-GR	443,665	1999	2.97	2.83	410.9	375.7
MSSP-GR	443,665	2000	2.45	2.86	234.5	378.2
MSSP-GR	443,665	2001	3.00	2.88	463.5	379.3
MSSP-GR	443,665	2002	2.96	2.89	378.1	379.1
MSSP-GR	443,665	2003	2.44	2.89	215.3	378.0
MSSP-GR	443,665	2004	2.90	2.89	359.2	375.8
MSSP-GR	443,665	2005	2.77	2.89	280.1	373.3
MSSP-GR	443,665	2006	2.61	2.89	210.1	371.9
MSSP-GR	443,665	2007	3.05	2.91	355.4	371.9
MSSP-GR	443,665	2008	3.18	2.94	564.0	373.8
MSSP-GR	443,665	2009	3.21	2.99	513.2	376.9
MSSP-GR	443,665	2010	3.29	3.04	601.3	381.3
MIZZ-HE	1,353,269	1980	0.83	0.98	50.5	94.9
MIZZ-HE	1,353,269	1981	0.83	1.00	59.5	96.0
MIZZ-HE	1,353,269	1982	1.06	1.01	119.3	97.3
MIZZ-HE	1,353,269	1983	1.16	1.03	135.4	98.8
MIZZ-HE	1,353,269	1984	1.20	1.06	155.9	100.5
MIZZ-HE	1,353,269	1985	1.18	1.08	129.0	101.7
MIZZ-HE	1,353,269	1986	1.24	1.09	135.2	102.5
MIZZ-HE	1,353,269	1987	1.19	1.11	113.8	103.5
MIZZ-HE	1,353,269	1988	0.91	1.12	53.6	104.2
MIZZ-HE	1,353,269	1989	0.92	1.14	42.2	105.2
MIZZ-HE	1,353,269	1990	1.14	1.16	95.7	106.6
MIZZ-HE	1,353,269	1991	1.10	1.18	65.5	107.7
MIZZ-HE	1,353,269	1992	1.21	1.19	91.6	108.4
MIZZ-HE	1,353,269	1993	1.28	1.20	214.8	108.7
MIZZ-HE	1,353,269	1994	1.20	1.19	112.9	108.4
MIZZ-HE	1,353,269	1995	1.25	1.19	151.9	108.1
MIZZ-HE	1,353,269	1996	1.29	1.19	128.8	107.9
MIZZ-HE	1,353,269	1997	1.25	1.19	140.1	108.0
MIZZ-HE	1,353,269	1998	1.26	1.20	150.1	108.5
MIZZ-HE	1,353,269	1999	1.23	1.21	140.6	109.6

Appendix 1.   Annual mean estimated and flow-normalized nitrate concentration and total annual estimated and flow-normalized nitrate flux from WRTDS for 1980 through 2010, by calendar years for eight sites in the Mississippi River Basin.—Continued

[WRTDS, weighted regressions on time, discharge, and season model; —, estimates not reported because too few samples were collected in that year; km², square kilometers; mg/L, milligrams per liter; kg/yr, kilograms per year]

Site	Drainage area, in km²	Calendar year	Annual mean estimated concentration, in mg/L	Annual mean flow-normalized concentration, in mg/L	Total annual estimated flux, in 10⁶ kg/yr	Total annual flow-normalized flux, in 10⁶ kg/yr
MIZZ-HE	1,353,269	2000	1.18	1.23	57.2	111.1
MIZZ-HE	1,353,269	2001	1.24	1.24	108.9	112.7
MIZZ-HE	1,353,269	2002	1.17	1.26	73.8	114.2
MIZZ-HE	1,353,269	2003	1.17	1.29	51.5	115.2
MIZZ-HE	1,353,269	2004	1.32	1.31	89.7	116.4
MIZZ-HE	1,353,269	2005	1.30	1.35	82.4	117.9
MIZZ-HE	1,353,269	2006	1.38	1.40	51.8	119.9
MIZZ-HE	1,353,269	2007	1.49	1.46	118.0	122.9
MIZZ-HE	1,353,269	2008	1.49	1.55	157.5	127.1
MIZZ-HE	1,353,269	2009	1.56	1.65	143.3	132.0
MIZZ-HE	1,353,269	2010	1.55	1.76	191.5	137.2
MSSP-TH	1,847,180	1980	1.62	1.96	250.0	496.3
MSSP-TH	1,847,180	1981	1.78	1.98	353.0	501.0
MSSP-TH	1,847,180	1982	2.25	2.00	669.0	505.3
MSSP-TH	1,847,180	1983	2.29	2.03	693.0	513.5
MSSP-TH	1,847,180	1984	2.37	2.06	724.0	519.0
MSSP-TH	1,847,180	1985	2.22	2.09	610.0	523.3
MSSP-TH	1,847,180	1986	2.44	2.12	680.0	526.6
MSSP-TH	1,847,180	1987	2.05	2.14	380.0	530.0
MSSP-TH	1,847,180	1988	1.69	2.17	242.0	533.9
MSSP-TH	1,847,180	1989	1.53	2.19	180.0	536.8
MSSP-TH	1,847,180	1990	2.16	2.20	497.0	537.9
MSSP-TH	1,847,180	1991	2.37	2.20	526.0	537.7
MSSP-TH	1,847,180	1992	2.24	2.21	457.0	537.4
MSSP-TH	1,847,180	1993	2.44	2.20	988.0	536.5
MSSP-TH	1,847,180	1994	2.28	2.19	493.0	533.4
MSSP-TH	1,847,180	1995	2.30	2.17	600.0	528.5
MSSP-TH	1,847,180	1996	2.26	2.15	543.0	521.6
MSSP-TH	1,847,180	1997	2.30	2.12	546.0	513.3
MSSP-TH	1,847,180	1998	2.34	2.10	658.0	506.4
MSSP-TH	1,847,180	1999	2.21	2.08	567.0	501.8
MSSP-TH	1,847,180	2000	1.81	2.07	276.0	500.4
MSSP-TH	1,847,180	2001	2.16	2.07	543.0	501.7
MSSP-TH	1,847,180	2002	2.07	2.09	435.0	505.1
MSSP-TH	1,847,180	2003	1.83	2.11	287.0	507.8
MSSP-TH	1,847,180	2004	2.18	2.13	458.0	509.8

Appendix 1.   Annual mean estimated and flow-normalized nitrate concentration and total annual estimated and flow-normalized nitrate flux from WRTDS for 1980 through 2010, by calendar years for eight sites in the Mississippi River Basin.—Continued

[WRTDS, weighted regressions on time, discharge, and season model; —, estimates not reported because too few samples were collected in that year; km², square kilometers; mg/L, milligrams per liter; kg/yr, kilograms per year]

Site	Drainage area, in km²	Calendar year	Annual mean estimated concentration, in mg/L	Annual mean flow-normalized concentration, in mg/L	Total annual estimated flux, in 10⁶ kg/yr	Total annual flow-normalized flux, in 10⁶ kg/yr
MSSP-TH	1,847,180	2005	2.01	2.15	360.0	511.1
MSSP-TH	1,847,180	2006	1.93	2.18	265.0	512.5
MSSP-TH	1,847,180	2007	2.33	2.21	498.0	515.8
MSSP-TH	1,847,180	2008	2.32	2.25	731.0	521.6
MSSP-TH	1,847,180	2009	2.32	2.30	678.0	529.6
MSSP-TH	1,847,180	2010	2.44	2.34	803.0	535.9
OHIO-GRCH	526,027	1980	1.02	0.99	290.0	310.0
OHIO-GRCH	526,027	1981	0.98	1.00	210.0	310.4
OHIO-GRCH	526,027	1982	1.03	1.00	338.0	310.4
OHIO-GRCH	526,027	1983	0.98	1.00	326.0	310.6
OHIO-GRCH	526,027	1984	1.01	1.00	337.0	311.5
OHIO-GRCH	526,027	1985	1.01	1.01	298.0	311.3
OHIO-GRCH	526,027	1986	0.98	1.02	222.0	313.3
OHIO-GRCH	526,027	1987	0.97	1.02	200.0	314.8
OHIO-GRCH	526,027	1988	0.89	1.03	173.0	316.2
OHIO-GRCH	526,027	1989	1.11	1.03	418.0	314.8
OHIO-GRCH	526,027	1990	1.07	1.02	379.0	313.8
OHIO-GRCH	526,027	1991	0.94	1.02	356.0	312.6
OHIO-GRCH	526,027	1992	1.08	1.02	240.0	312.1
OHIO-GRCH	526,027	1993	0.97	1.01	298.0	309.9
OHIO-GRCH	526,027	1994	0.99	1.02	361.0	309.6
OHIO-GRCH	526,027	1995	1.03	1.02	271.0	310.2
OHIO-GRCH	526,027	1996	1.07	1.03	424.0	311.7
OHIO-GRCH	526,027	1997	1.03	1.03	347.0	311.7
OHIO-GRCH	526,027	1998	1.04	1.04	351.0	312.7
OHIO-GRCH	526,027	1999	1.00	1.04	235.0	313.2
OHIO-GRCH	526,027	2000	1.05	1.04	202.0	314.3
OHIO-GRCH	526,027	2001	1.06	1.04	237.0	313.3
OHIO-GRCH	526,027	2002	1.01	1.04	307.0	312.8
OHIO-GRCH	526,027	2003	1.09	1.04	390.0	311.9
OHIO-GRCH	526,027	2004	1.08	1.03	380.0	311.2
OHIO-GRCH	526,027	2005	0.99	1.02	273.0	308.4
OHIO-GRCH	526,027	2006	1.06	1.02	267.0	306.2
OHIO-GRCH	526,027	2007	0.91	1.01	233.0	303.3
OHIO-GRCH	526,027	2008	0.93	1.00	308.0	301.2
OHIO-GRCH	526,027	2009	1.03	0.99	303.0	297.4

Appendix 1.   Annual mean estimated and flow-normalized nitrate concentration and total annual estimated and flow-normalized nitrate flux from WRTDS for 1980 through 2010, by calendar years for eight sites in the Mississippi River Basin.—Continued

[WRTDS, weighted regressions on time, discharge, and season model; —, estimates not reported because too few samples were collected in that year; km², square kilometers; mg/L, milligrams per liter; kg/yr, kilograms per year]

Site	Drainage area, in km²	Calendar year	Annual mean estimated concentration, in mg/L	Annual mean flow-normalized concentration, in mg/L	Total annual estimated flux, in 10⁶ kg/yr	Total annual flow-normalized flux, in 10⁶ kg/yr
OHIO-GRCH	526,027	2010	0.95	0.98	247.0	294.1
MSSP-OUT	2,914,514	1980	1.25	1.26	667.0	818.5
MSSP-OUT	2,914,514	1981	1.25	1.30	578.0	854.1
MSSP-OUT	2,914,514	1982	1.40	1.33	936.0	884.5
MSSP-OUT	2,914,514	1983	1.38	1.35	1,180.0	902.4
MSSP-OUT	2,914,514	1984	1.43	1.36	1,099.0	907.7
MSSP-OUT	2,914,514	1985	1.39	1.37	987.0	909.9
MSSP-OUT	2,914,514	1986	1.44	1.37	890.0	910.8
MSSP-OUT	2,914,514	1987	1.23	1.37	623.0	909.2
MSSP-OUT	2,914,514	1988	0.96	1.36	527.0	904.0
MSSP-OUT	2,914,514	1989	1.38	1.35	934.0	893.4
MSSP-OUT	2,914,514	1990	1.39	1.33	1,005.0	879.7
MSSP-OUT	2,914,514	1991	1.26	1.32	967.0	868.1
MSSP-OUT	2,914,514	1992	1.36	1.31	732.0	858.9
MSSP-OUT	2,914,514	1993	1.49	1.31	1,356.0	855.4
MSSP-OUT	2,914,514	1994	1.30	1.31	896.0	858.1
MSSP-OUT	2,914,514	1995	1.35	1.31	871.0	862.2
MSSP-OUT	2,914,514	1996	1.38	1.32	988.0	865.3
MSSP-OUT	2,914,514	1997	1.32	1.32	966.0	864.2
MSSP-OUT	2,914,514	1998	1.36	1.32	1,023.0	860.4
MSSP-OUT	2,914,514	1999	1.29	1.32	839.0	855.0
MSSP-OUT	2,914,514	2000	1.28	1.31	557.0	852.2
MSSP-OUT	2,914,514	2001	1.32	1.31	776.0	851.8
MSSP-OUT	2,914,514	2002	1.31	1.32	860.0	853.7
MSSP-OUT	2,914,514	2003	1.36	1.33	845.0	857.3
MSSP-OUT	2,914,514	2004	1.37	1.34	957.0	860.9
MSSP-OUT	2,914,514	2005	1.35	1.35	742.0	863.3
MSSP-OUT	2,914,514	2006	1.39	1.36	604.0	866.7
MSSP-OUT	2,914,514	2007	1.42	1.37	787.0	874.8
MSSP-OUT	2,914,514	2008	1.33	1.40	1,044.0	890.4
MSSP-OUT	2,914,514	2009	1.40	1.43	1,051.0	911.9
MSSP-OUT	2,914,514	2010	1.49	1.47	1,054.0	937.0

Appendix 2. Spring (April, May, and June) mean estimated and flow-normalized nitrate concentration and total spring estimated and flow-normalized nitrate flux from WRTDS for 1980 through 2010 for eight sites in the Mississippi River Basin.

[WRTDS, weighted regressions on time, discharge, and season model; —, estimates not reported because too few samples were collected in that year; km², square kilometers; mg/L, milligrams per liter; kg/spr, kilograms per 91-day spring period (April, May, and June)]

Site	Drainage area, in km²	Calendar year	Spring mean estimated concentration, in mg/L	Spring mean flow-normalized concentration, in mg/L	Total spring estimated flux, in 10⁶ kg/spr	Total spring flow-normalized flux, in 10⁶ kg/spr
MSSP-CL	221,703	1980	0.77	1.25	11.6	30.0
MSSP-CL	221,703	1981	0.86	1.35	12.3	32.1
MSSP-CL	221,703	1982	1.64	1.40	40.2	33.3
MSSP-CL	221,703	1983	1.54	1.47	32.8	34.6
MSSP-CL	221,703	1984	1.80	1.51	41.2	35.7
MSSP-CL	221,703	1985	1.57	1.59	32.1	37.4
MSSP-CL	221,703	1986	2.25	1.66	63.5	38.5
MSSP-CL	221,703	1987	0.89	1.73	9.2	39.7
MSSP-CL	221,703	1988	—	—	—	—
MSSP-CL	221,703	1989	—	—	—	—
MSSP-CL	221,703	1990	—	—	—	—
MSSP-CL	221,703	1991	2.69	2.09	65.2	46.3
MSSP-CL	221,703	1992	1.80	2.04	34.6	44.8
MSSP-CL	221,703	1993	2.79	1.98	90.2	43.4
MSSP-CL	221,703	1994	—	—	—	—
MSSP-CL	221,703	1995	2.07	1.86	43.0	40.7
MSSP-CL	221,703	1996	2.23	1.79	56.1	39.1
MSSP-CL	221,703	1997	1.86	1.73	49.7	37.6
MSSP-CL	221,703	1998	1.70	1.69	35.7	36.4
MSSP-CL	221,703	1999	1.88	1.66	42.2	35.5
MSSP-CL	221,703	2000	1.48	1.63	24.5	34.8
MSSP-CL	221,703	2001	2.22	1.66	77.5	35.4
MSSP-CL	221,703	2002	1.88	1.72	39.8	36.9
MSSP-CL	221,703	2003	1.75	1.79	32.9	38.3
MSSP-CL	221,703	2004	2.02	1.84	49.0	39.3
MSSP-CL	221,703	2005	1.87	1.88	32.8	40.1
MSSP-CL	221,703	2006	1.90	1.92	34.7	40.8
MSSP-CL	221,703	2007	1.76	1.97	27.7	41.8
MSSP-CL	221,703	2008	2.50	2.05	73.4	43.4
MSSP-CL	221,703	2009	1.90	2.14	27.6	45.1
MSSP-CL	221,703	2010	2.12	2.21	36.4	46.7
IOWA-WAP	32,375	1980	4.67	6.00	10.1	26.7
IOWA-WAP	32,375	1981	4.19	5.90	7.1	26.1
IOWA-WAP	32,375	1982	6.98	5.95	30.5	26.3
IOWA-WAP	32,375	1983	6.98	5.87	36.2	25.9
IOWA-WAP	32,375	1984	7.18	5.86	35.3	25.8

Appendix 2.   Spring (April, May, and June) mean estimated and flow-normalized nitrate concentration and total spring estimated and flow-normalized nitrate flux from WRTDS for 1980 through 2010 for eight sites in the Mississippi River Basin.—Continued

[WRTDS, weighted regressions on time, discharge, and season model; —, estimates not reported because too few samples were collected in that year; km², square kilometers; mg/L, milligrams per liter; kg/spr, kilograms per 91-day spring period (April, May, and June)]

Site	Drainage area, in km²	Calendar year	Spring mean estimated concentration, in mg/L	Spring mean flow-normalized concentration, in mg/L	Total spring estimated flux, in 10⁶ kg/spr	Total spring flow-normalized flux, in 10⁶ kg/spr
IOWA-WAP	32,375	1985	3.50	5.86	5.3	25.9
IOWA-WAP	32,375	1986	6.60	5.87	26.9	26.1
IOWA-WAP	32,375	1987	4.30	5.90	8.1	26.4
IOWA-WAP	32,375	1988	2.78	5.97	3.3	26.9
IOWA-WAP	32,375	1989	1.51	6.06	1.1	27.5
IOWA-WAP	32,375	1990	5.40	6.12	23.4	27.8
IOWA-WAP	32,375	1991	8.17	6.17	53.2	28.1
IOWA-WAP	32,375	1992	5.40	6.20	15.6	28.2
IOWA-WAP	32,375	1993	8.42	6.22	71.3	28.3
IOWA-WAP	32,375	1994	4.58	6.23	8.7	28.3
IOWA-WAP	32,375	1995	7.67	6.23	36.7	28.2
IOWA-WAP	32,375	1996	6.36	6.23	28.3	28.1
IOWA-WAP	32,375	1997	6.29	6.24	18.8	28.2
IOWA-WAP	32,375	1998	7.50	6.35	39.8	28.5
IOWA-WAP	32,375	1999	8.09	6.50	50.7	29.0
IOWA-WAP	32,375	2000	5.21	6.66	16.9	29.4
IOWA-WAP	32,375	2001	8.36	6.77	49.8	29.5
IOWA-WAP	32,375	2002	6.25	6.83	14.0	29.4
IOWA-WAP	32,375	2003	6.31	6.85	17.6	29.0
IOWA-WAP	32,375	2004	6.77	6.86	29.8	28.7
IOWA-WAP	32,375	2005	6.92	6.85	17.9	28.5
IOWA-WAP	32,375	2006	7.01	6.82	20.3	28.2
IOWA-WAP	32,375	2007	7.69	6.76	38.8	27.9
IOWA-WAP	32,375	2008	7.25	6.66	69.5	27.4
IOWA-WAP	32,375	2009	7.16	6.55	30.6	26.9
IOWA-WAP	32,375	2010	7.22	6.42	38.5	26.4
ILLI-VC	69,264	1980	4.91	4.90	38.2	44.1
ILLI-VC	69,264	1981	5.20	4.88	53.6	44.0
ILLI-VC	69,264	1982	5.13	4.87	53.8	44.0
ILLI-VC	69,264	1983	5.34	4.89	71.5	44.3
ILLI-VC	69,264	1984	5.38	4.90	57.8	44.4
ILLI-VC	69,264	1985	4.43	4.91	31.4	44.5
ILLI-VC	69,264	1986	4.62	4.93	23.9	44.6
ILLI-VC	69,264	1987	4.26	4.93	17.6	44.5
ILLI-VC	69,264	1988	3.56	4.92	16.1	44.3
ILLI-VC	69,264	1989	4.05	4.90	14.1	44.1

Appendix 2. Spring (April, May, and June) mean estimated and flow-normalized nitrate concentration and total spring estimated and flow-normalized nitrate flux from WRTDS for 1980 through 2010 for eight sites in the Mississippi River Basin.—Continued

[WRTDS, weighted regressions on time, discharge, and season model; —, estimates not reported because too few samples were collected in that year; km², square kilometers; mg/L, milligrams per liter; kg/spr, kilograms per 91-day spring period (April, May, and June)]

Site	Drainage area, in km²	Calendar year	Spring mean estimated concentration, in mg/L	Spring mean flow-normalized concentration, in mg/L	Total spring estimated flux, in 10⁶ kg/spr	Total spring flow-normalized flux, in 10⁶ kg/spr
ILLI-VC	69,264	1990	5.34	4.88	52.1	43.8
ILLI-VC	69,264	1991	5.20	4.87	54.3	43.7
ILLI-VC	69,264	1992	4.01	4.88	15.9	43.8
ILLI-VC	69,264	1993	5.23	4.91	61.4	44.1
ILLI-VC	69,264	1994	4.90	4.96	39.9	44.6
ILLI-VC	69,264	1995	5.59	5.01	75.7	45.1
ILLI-VC	69,264	1996	5.34	5.05	57.4	45.7
ILLI-VC	69,264	1997	4.89	5.10	28.6	46.3
ILLI-VC	69,264	1998	5.94	5.16	74.9	47.1
ILLI-VC	69,264	1999	5.74	5.25	56.5	48.1
ILLI-VC	69,264	2000	4.52	5.31	21.4	48.7
ILLI-VC	69,264	2001	5.35	5.30	38.2	48.0
ILLI-VC	69,264	2002	5.77	5.25	73.5	47.0
ILLI-VC	69,264	2003	4.76	5.19	23.7	46.2
ILLI-VC	69,264	2004	5.11	5.12	38.1	45.6
ILLI-VC	69,264	2005	4.20	5.04	15.3	45.0
ILLI-VC	69,264	2006	4.63	4.92	24.4	44.1
ILLI-VC	69,264	2007	4.75	4.76	39.2	42.9
ILLI-VC	69,264	2008	4.85	4.57	45.5	41.2
ILLI-VC	69,264	2009	5.06	4.37	81.0	39.5
ILLI-VC	69,264	2010	4.49	4.15	50.0	37.5
MSSP-GR	443,665	1980	2.98	3.19	89.5	145.7
MSSP-GR	443,665	1981	3.06	3.20	106.1	146.7
MSSP-GR	443,665	1982	3.46	3.20	178.9	146.9
MSSP-GR	443,665	1983	3.54	3.18	212.1	146.8
MSSP-GR	443,665	1984	3.51	3.16	188.4	146.2
MSSP-GR	443,665	1985	2.89	3.13	100.6	145.6
MSSP-GR	443,665	1986	3.40	3.10	167.7	145.2
MSSP-GR	443,665	1987	2.44	3.07	58.9	144.4
MSSP-GR	443,665	1988	1.68	3.05	36.9	144.6
MSSP-GR	443,665	1989	1.98	3.06	40.7	145.7
MSSP-GR	443,665	1990	2.99	3.07	124.1	146.4
MSSP-GR	443,665	1991	3.60	3.09	191.4	147.6
MSSP-GR	443,665	1992	2.49	3.12	81.0	148.7
MSSP-GR	443,665	1993	3.84	3.17	262.4	150.4
MSSP-GR	443,665	1994	3.15	3.24	113.0	153.2

Appendix 2. Spring (April, May, and June) mean estimated and flow-normalized nitrate concentration and total spring estimated and flow-normalized nitrate flux from WRTDS for 1980 through 2010 for eight sites in the Mississippi River Basin.—Continued

[WRTDS, weighted regressions on time, discharge, and season model; —, estimates not reported because too few samples were collected in that year; km², square kilometers; mg/L, milligrams per liter; kg/spr, kilograms per 91-day spring period (April, May, and June)]

Site	Drainage area, in km²	Calendar year	Spring mean estimated concentration, in mg/L	Spring mean flow-normalized concentration, in mg/L	Total spring estimated flux, in 10⁶ kg/spr	Total spring flow-normalized flux, in 10⁶ kg/spr
MSSP-GR	443,665	1995	—	—	—	—
MSSP-GR	443,665	1996	3.71	3.39	192.9	159.5
MSSP-GR	443,665	1997	3.55	3.45	145.4	162.0
MSSP-GR	443,665	1998	3.90	3.52	205.8	164.2
MSSP-GR	443,665	1999	3.94	3.59	207.1	166.4
MSSP-GR	443,665	2000	3.16	3.66	102.3	168.8
MSSP-GR	443,665	2001	4.15	3.73	261.8	171.1
MSSP-GR	443,665	2002	4.02	3.77	220.7	172.2
MSSP-GR	443,665	2003	3.63	3.79	121.3	172.2
MSSP-GR	443,665	2004	3.71	3.78	169.3	171.1
MSSP-GR	443,665	2005	3.76	3.75	117.6	169.3
MSSP-GR	443,665	2006	3.68	3.74	116.0	168.3
MSSP-GR	443,665	2007	3.89	3.75	151.5	167.9
MSSP-GR	443,665	2008	3.89	3.78	264.0	168.5
MSSP-GR	443,665	2009	3.96	3.84	218.3	170.1
MSSP-GR	443,665	2010	3.97	3.91	230.5	172.5
MIZZ-HE	1,353,269	1980	1.17	1.32	25.3	41.4
MIZZ-HE	1,353,269	1981	1.07	1.33	24.3	42.1
MIZZ-HE	1,353,269	1982	1.26	1.34	42.7	42.7
MIZZ-HE	1,353,269	1983	1.69	1.37	74.7	43.5
MIZZ-HE	1,353,269	1984	1.66	1.41	81.1	44.4
MIZZ-HE	1,353,269	1985	1.57	1.43	44.7	44.9
MIZZ-HE	1,353,269	1986	1.61	1.46	44.3	45.4
MIZZ-HE	1,353,269	1987	1.69	1.48	51.2	45.9
MIZZ-HE	1,353,269	1988	1.13	1.51	19.8	46.3
MIZZ-HE	1,353,269	1989	1.04	1.55	13.3	46.8
MIZZ-HE	1,353,269	1990	1.65	1.59	56.0	47.5
MIZZ-HE	1,353,269	1991	1.71	1.63	37.9	48.0
MIZZ-HE	1,353,269	1992	1.61	1.66	26.6	48.4
MIZZ-HE	1,353,269	1993	1.67	1.68	70.0	48.7
MIZZ-HE	1,353,269	1994	1.74	1.67	60.5	48.7
MIZZ-HE	1,353,269	1995	1.59	1.66	82.6	48.4
MIZZ-HE	1,353,269	1996	1.67	1.65	59.7	48.3
MIZZ-HE	1,353,269	1997	1.70	1.64	63.3	48.2
MIZZ-HE	1,353,269	1998	1.68	1.65	56.6	48.5
MIZZ-HE	1,353,269	1999	1.71	1.66	72.8	48.8

Appendix 2.   Spring (April, May, and June) mean estimated and flow-normalized nitrate concentration and total spring estimated and flow-normalized nitrate flux from WRTDS for 1980 through 2010 for eight sites in the Mississippi River Basin.—Continued

[WRTDS, weighted regressions on time, discharge, and season model; —, estimates not reported because too few samples were collected in that year; km², square kilometers; mg/L, milligrams per liter; kg/spr, kilograms per 91-day spring period (April, May, and June)]

Site	Drainage area, in km²	Calendar year	Spring mean estimated concentration, in mg/L	Spring mean flow-normalized concentration, in mg/L	Total spring estimated flux, in 10⁶ kg/spr	Total spring flow-normalized flux, in 10⁶ kg/spr
MIZZ-HE	1,353,269	2000	1.61	1.66	21.7	49.3
MIZZ-HE	1,353,269	2001	1.69	1.67	56.1	50.0
MIZZ-HE	1,353,269	2002	1.67	1.70	45.6	50.8
MIZZ-HE	1,353,269	2003	1.62	1.72	23.6	51.4
MIZZ-HE	1,353,269	2004	1.71	1.75	34.7	51.7
MIZZ-HE	1,353,269	2005	1.77	1.78	35.0	52.1
MIZZ-HE	1,353,269	2006	1.93	1.82	24.5	52.4
MIZZ-HE	1,353,269	2007	1.79	1.88	58.7	53.1
MIZZ-HE	1,353,269	2008	1.76	1.95	70.2	53.9
MIZZ-HE	1,353,269	2009	1.79	2.03	62.3	54.9
MIZZ-HE	1,353,269	2010	1.74	2.09	75.9	55.2
MSSP-TH	1,847,180	1980	2.03	2.54	114.3	221.2
MSSP-TH	1,847,180	1981	2.16	2.59	133.5	226.0
MSSP-TH	1,847,180	1982	2.94	2.64	262.9	230.6
MSSP-TH	1,847,180	1983	3.23	2.71	367.1	236.6
MSSP-TH	1,847,180	1984	3.30	2.76	352.9	238.8
MSSP-TH	1,847,180	1985	2.87	2.79	210.3	238.9
MSSP-TH	1,847,180	1986	3.07	2.82	241.7	238.0
MSSP-TH	1,847,180	1987	2.55	2.84	146.4	237.3
MSSP-TH	1,847,180	1988	2.05	2.86	82.5	237.6
MSSP-TH	1,847,180	1989	2.12	2.90	80.9	238.3
MSSP-TH	1,847,180	1990	2.97	2.92	241.7	238.5
MSSP-TH	1,847,180	1991	3.39	2.93	271.1	237.7
MSSP-TH	1,847,180	1992	2.57	2.92	131.6	236.9
MSSP-TH	1,847,180	1993	2.94	2.91	331.0	236.1
MSSP-TH	1,847,180	1994	3.00	2.88	225.1	234.3
MSSP-TH	1,847,180	1995	2.78	2.85	301.3	231.8
MSSP-TH	1,847,180	1996	2.90	2.80	276.3	228.3
MSSP-TH	1,847,180	1997	3.03	2.76	238.7	224.6
MSSP-TH	1,847,180	1998	2.92	2.72	268.6	221.5
MSSP-TH	1,847,180	1999	2.93	2.69	279.7	219.6
MSSP-TH	1,847,180	2000	2.24	2.68	107.2	219.4
MSSP-TH	1,847,180	2001	2.98	2.70	283.1	221.4
MSSP-TH	1,847,180	2002	2.84	2.74	251.3	224.1
MSSP-TH	1,847,180	2003	2.66	2.76	153.5	225.7
MSSP-TH	1,847,180	2004	2.82	2.78	201.6	226.0

Appendix 2.   Spring (April, May, and June) mean estimated and flow-normalized nitrate concentration and total spring estimated and flow-normalized nitrate flux from WRTDS for 1980 through 2010 for eight sites in the Mississippi River Basin.—Continued

[WRTDS, weighted regressions on time, discharge, and season model; —, estimates not reported because too few samples were collected in that year; km^2, square kilometers; mg/L, milligrams per liter; kg/spr, kilograms per 91-day spring period (April, May, and June)]

Site	Drainage area, in km^2	Calendar year	Spring mean estimated concentration, in mg/L	Spring mean flow-normalized concentration, in mg/L	Total spring estimated flux, in 10^6 kg/spr	Total spring flow-normalized flux, in 10^6 kg/spr
MSSP-TH	1,847,180	2005	2.82	2.79	147.5	225.5
MSSP-TH	1,847,180	2006	2.72	2.79	132.8	224.7
MSSP-TH	1,847,180	2007	3.00	2.82	231.7	225.4
MSSP-TH	1,847,180	2008	2.74	2.87	327.4	227.9
MSSP-TH	1,847,180	2009	2.91	2.94	295.5	232.1
MSSP-TH	1,847,180	2010	2.90	2.96	305.4	233.4
OHIO-GRCH	526,027	1980	1.21	1.12	107.4	94.1
OHIO-GRCH	526,027	1981	1.18	1.15	93.7	95.0
OHIO-GRCH	526,027	1982	1.22	1.16	75.8	95.5
OHIO-GRCH	526,027	1983	1.12	1.17	145.9	95.7
OHIO-GRCH	526,027	1984	1.16	1.18	141.4	96.0
OHIO-GRCH	526,027	1985	1.28	1.19	71.1	96.1
OHIO-GRCH	526,027	1986	1.18	1.21	39.3	97.4
OHIO-GRCH	526,027	1987	1.24	1.23	67.9	99.0
OHIO-GRCH	526,027	1988	1.08	1.24	41.2	100.1
OHIO-GRCH	526,027	1989	1.26	1.24	134.4	100.4
OHIO-GRCH	526,027	1990	1.29	1.23	104.7	100.5
OHIO-GRCH	526,027	1991	1.19	1.23	88.8	100.8
OHIO-GRCH	526,027	1992	1.27	1.23	58.0	100.9
OHIO-GRCH	526,027	1993	1.18	1.24	89.4	101.2
OHIO-GRCH	526,027	1994	1.17	1.25	119.4	101.8
OHIO-GRCH	526,027	1995	1.33	1.27	102.5	102.8
OHIO-GRCH	526,027	1996	1.32	1.29	170.3	103.6
OHIO-GRCH	526,027	1997	1.37	1.30	123.4	104.1
OHIO-GRCH	526,027	1998	1.32	1.31	157.2	104.5
OHIO-GRCH	526,027	1999	1.31	1.31	54.8	104.4
OHIO-GRCH	526,027	2000	1.33	1.31	71.5	104.1
OHIO-GRCH	526,027	2001	1.34	1.30	62.7	103.0
OHIO-GRCH	526,027	2002	1.23	1.28	128.4	101.8
OHIO-GRCH	526,027	2003	1.27	1.26	129.9	100.5
OHIO-GRCH	526,027	2004	1.27	1.25	107.3	99.3
OHIO-GRCH	526,027	2005	1.21	1.23	71.4	98.0
OHIO-GRCH	526,027	2006	1.25	1.22	63.0	96.9
OHIO-GRCH	526,027	2007	1.10	1.20	54.1	95.9
OHIO-GRCH	526,027	2008	1.21	1.20	118.8	95.4
OHIO-GRCH	526,027	2009	1.22	1.19	110.3	95.0

Appendix 2.   Spring (April, May, and June) mean estimated and flow-normalized nitrate concentration and total spring estimated and flow-normalized nitrate flux from WRTDS for 1980 through 2010 for eight sites in the Mississippi River Basin.—Continued

[WRTDS, weighted regressions on time, discharge, and season model; —, estimates not reported because too few samples were collected in that year; km², square kilometers; mg/L, milligrams per liter; kg/spr, kilograms per 91-day spring period (April, May, and June)]

Site	Drainage area, in km²	Calendar year	Spring mean estimated concentration, in mg/L	Spring mean flow-normalized concentration, in mg/L	Total spring estimated flux, in 10⁶ kg/spr	Total spring flow-normalized flux, in 10⁶ kg/spr
OHIO-GRCH	526,027	2010	1.23	1.19	89.4	94.8
MSSP-OUT	2,914,514	1980	1.59	1.54	311.5	326.2
MSSP-OUT	2,914,514	1981	1.54	1.62	234.3	346.2
MSSP-OUT	2,914,514	1982	1.73	1.68	340.4	363.3
MSSP-OUT	2,914,514	1983	1.79	1.71	591.1	372.1
MSSP-OUT	2,914,514	1984	1.80	1.72	523.5	373.2
MSSP-OUT	2,914,514	1985	1.76	1.72	338.1	373.0
MSSP-OUT	2,914,514	1986	1.76	1.72	281.6	372.4
MSSP-OUT	2,914,514	1987	1.60	1.71	240.1	371.0
MSSP-OUT	2,914,514	1988	1.18	1.70	161.4	368.5
MSSP-OUT	2,914,514	1989	1.73	1.68	338.8	363.1
MSSP-OUT	2,914,514	1990	1.71	1.65	426.8	356.3
MSSP-OUT	2,914,514	1991	1.68	1.63	400.0	350.1
MSSP-OUT	2,914,514	1992	1.57	1.61	211.7	344.6
MSSP-OUT	2,914,514	1993	1.63	1.61	458.6	343.2
MSSP-OUT	2,914,514	1994	1.63	1.63	383.8	346.0
MSSP-OUT	2,914,514	1995	1.67	1.65	383.6	350.1
MSSP-OUT	2,914,514	1996	1.68	1.67	422.4	352.9
MSSP-OUT	2,914,514	1997	1.71	1.68	410.0	352.8
MSSP-OUT	2,914,514	1998	1.68	1.68	421.9	350.8
MSSP-OUT	2,914,514	1999	1.74	1.68	350.6	349.3
MSSP-OUT	2,914,514	2000	1.74	1.69	227.2	350.1
MSSP-OUT	2,914,514	2001	1.79	1.71	314.2	352.4
MSSP-OUT	2,914,514	2002	1.67	1.73	429.0	354.2
MSSP-OUT	2,914,514	2003	1.77	1.74	348.0	354.7
MSSP-OUT	2,914,514	2004	1.81	1.74	346.4	354.2
MSSP-OUT	2,914,514	2005	1.92	1.75	272.6	352.9
MSSP-OUT	2,914,514	2006	1.97	1.75	237.0	351.7
MSSP-OUT	2,914,514	2007	1.93	1.77	310.1	353.3
MSSP-OUT	2,914,514	2008	1.45	1.81	451.3	359.3
MSSP-OUT	2,914,514	2009	1.67	1.86	420.2	369.5
MSSP-OUT	2,914,514	2010	1.82	1.92	410.3	380.3